THE BOOK OF MAGIC SQUARES

JAIN

VOLUME 2

2nd EDITION
YEAR 2000

THE BOOK OF MAGIC SQUARES
VOLUME 2

ISBN: 978-1-925834-32-1

1st Edition (1996)   2nd Edition (2000)

Self Published by F.R.E.E.D.O.M.S. an acronym for:

For **R**esearch **E**xpressing **E**ssential **D**ata **O**f **M**agic **S**quares

For Correspondence, contact:

JAIN

c/o MAGIC SQUARE STUDIO

777 LEFT BANK RD, MULLUMBIMBY CREEK, Northern N.S.W. Australia, 2482.

Phone: 61 2 66 844409

email: jain@jainmathemagics.com

website: www.jainmathemagics.com

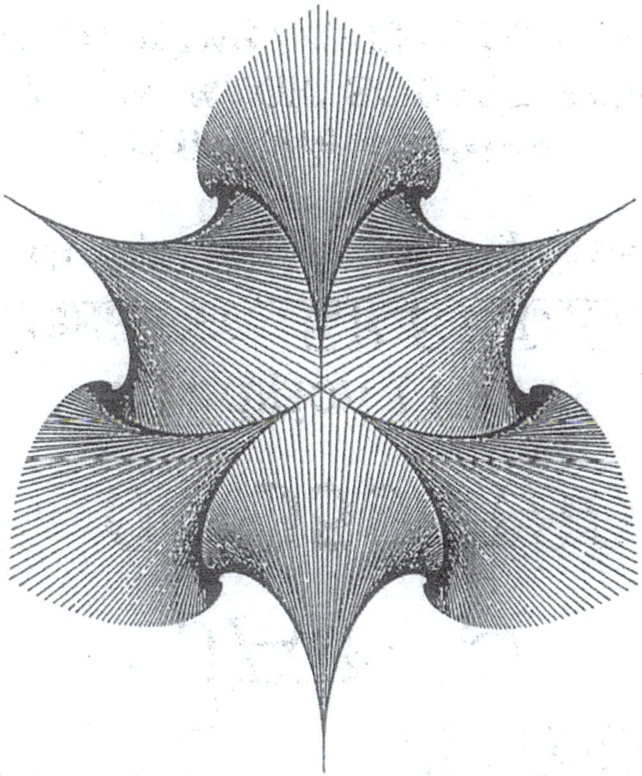

Crest

Courtesy of California Computer Products, Inc.

THE BOOK OF MAGIC SQUARES

Volume 1
2nd Edition Year 2000
As gathered by JAIN ©

Previously known as: The Book of Harmony Squares
Volume One
JAIN
1990

1st Edition: Printed and bound in Australia by Gympie Times Printing. Limited Edition of 1,000 copies, A3 size but in The Phi Ratio.

ISBN: 1ST EDITION 0 - 9594180 - 1 - 6

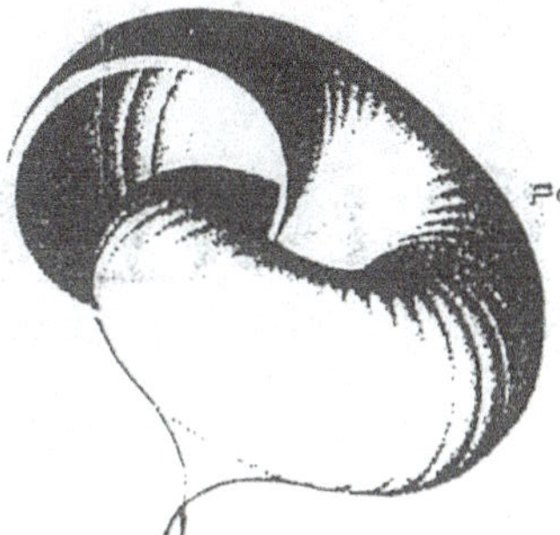

F.R.E.E.D.O.M.S.
For Research Expressing Essential Data Of Magic Squares

ADDRESS:
MAGIC SQUARE STUDIO
777 LEFT BANK RD MULLUMBIMBY CREEK
2482 N.S.W AUSTRALIA.
PHONE:
0011 61 2 66 844409

THE BOOK OF MAGIC SQUARES

JAIN 2ND EDITION Year 2000 Volume 2

ISBN : 0 — 9594180 – 6 – 7

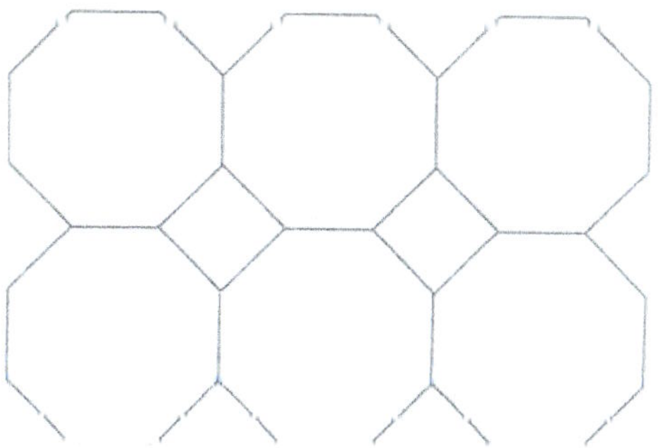

THIS IS MADE IN LOVE FOR LOVE

Please feel free to bootleg, reproduce, rip-off, any portion (ie chapter) of this book, whether it be by photocopying, transmitted in any form or by any means or stored in a retrieval system, electronic, mechanical, whatever. Really, there is no Copyright. Everything is Public Domain. How can any one person say "This is mine". How can you put your name to anything! All belongs to the Universal Archives, the Akashic Records, the Vibrational Library. As with Knowledge, Theories, Documents & Inventions, so it is with Healing – it comes from the God I AM Source withIN; there is no cost for it. I ask only for acknowledgement as an author, composer, inventor, sacred geometrician as a means for networking what my passion is. I reserve the right to publish this book in its entirety. If you choose to reprint any portion or chapter kindly acknowledge the front cover details & make prior contact with me for my own records. But that is all. No copyright, no name, no ego. Sometimes, independently 3 or 4 people in different countries tap or tune into the same idea or invention or concept. e.g: There is a Prof. John Searl of England who has created clean, non-polluting Free Energy systems, generators, levity discs, spacecrafts etc created via specific combinations of powdered metals relating to the Periodic Table of Elements & rings & rollers within magnetic field geometries & all the mathematics is determined specifically from MAGIC SQUAREOLOGY! So tap into these squares, give them to your CHILDREN & thus access FREE pure ENERGY. Magic Squares equal unlimited Free Energy! All information is for sharing.

PAGE A3

PAGE B

EDITORIAL

The Process Is The Goal.

"Success is a Journey not a destination".
B. Sweetland.

"CHANGES THAT HAVE BEEN MADE".

This book is half of a larger body of work that had the same title. The first edition was a 290 page compendium. Many readers have been overwhelmed by such a large book + it is for this reason that it now exists as two smaller books: Volumes 2 + 3. This book deals only with Magic Squares of 8, 9 + 10. It is a sequel to my first book "THE BOOK OF HARMONY SQUARES" Vol 1 1990. This latter title has also been changed to "THE BOOK OF MAGIC SQUARES" Vol 1. The contents are the same, only the title is difrnt. The reason for this change, from "Harmony Sqs." to "Magic Sqs." is that 20 yrs ago some Christian Schools objected to the word "Magic" when I was teaching there. I feel now that I choose not to change or interfere with ancient titles, and I am proud to introduce such a body of work as "Magic Squares", despite the ongoing controversy. Currently the State of Queensland (Department of Education run by Fundamentalist Christians) has formally banned all my works, presence + books in their State, objecting to this word + its superstitious associations. Fortunately or unfortunately, Magic Square Artforms produce incredible atomic patterns formed from combinations of Christian Crosses + Eastern Swaztikas! This was revealed in Vol 1, page 110. Besides this, Magic Squares are enjoyed by all other State schools in Australia. If you are a Maths Teacher in Qld you can support me by employing me as an Authorized School Performer, my unique theatrical show being known as "MATHEMAGICS".

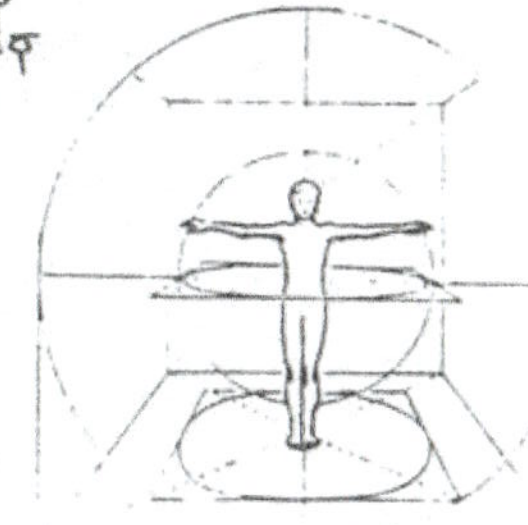

This book of Magic Squares
* Contains over **77** unique and original Yantras or constructional activities to draw, merely by Joining-The-Dots.
* Contains **43** original fine pen drawings by Jain. (These are identified by a small signature near the drawing).

This Compendium contains :

- ◊ **160** pages.
- ◊ **77** original drawings/ paintings
- ◊ **43** geometric/M.Sq. Activities to draw.

PAGE D

THE TWO MULES
A FABLE for the NATIONS

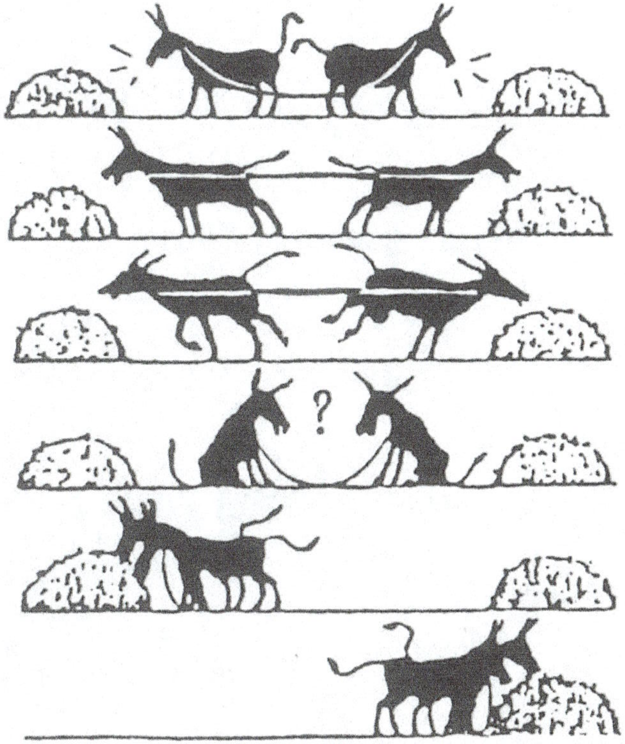

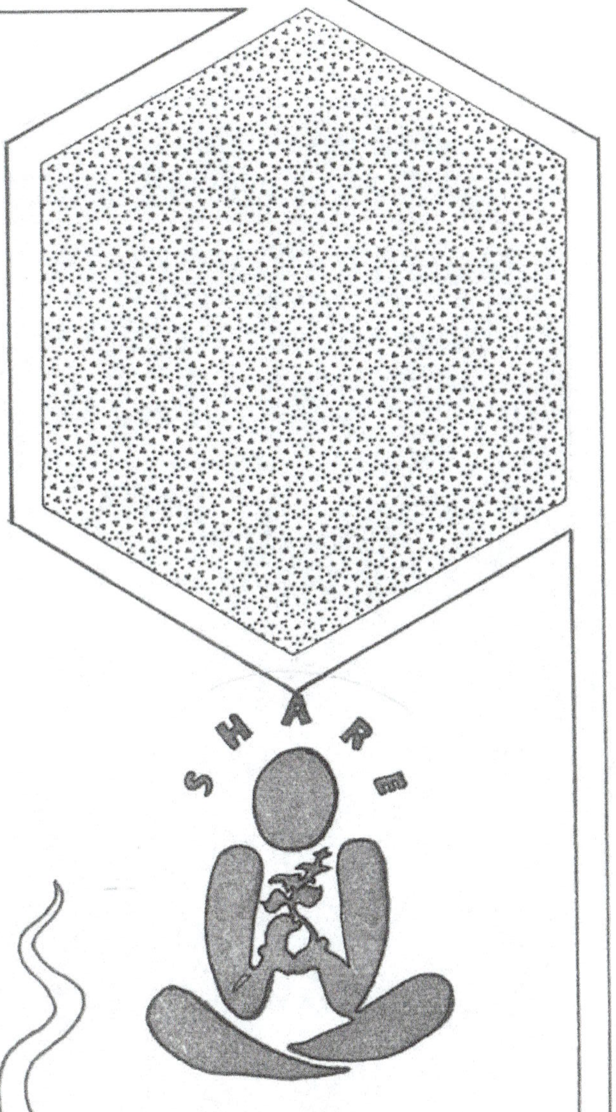

"The greatest divine revelation on earth is the evolving human being".

Rudolf Steiner

God-dess-ence
I Am.
This Book
Belongs to:
To Live Life
To Love Life
To Live Love.
Page 1

Page 2

Dedication to Asha and Ming-kah Sun.

"To my Guides & Spirits & all those who are assisting me in my evolutionary journey on this Earth;
it is my intention that I am safe always in the things that I do;
it is my intention that I receive Love & that I give Love in the things that I do;
it is my intention that I have a darn good time, that I am provided for with prosperity according to my needs;
it is my intention that I do not get overly enamoured with the physical world.
In Joy, in Safety & in Harmony I step into the Unknown."

(A Prayer for the Pleiadian Children, by Barbara Marciniak).

Acknowledgements

I would like to thank :

a) The Montessori School in the Adelaide Hills '86
b) Apollo Bay High School, Victoria '86
c) Didi of Murwillumbah N.S.W. Kangia Steiner School 1991
d) Julie Meana, Redcliffe Catholic, Brisbane '92
e) Ken Graham, Maths Master of Smithfield High School, Cairns '93
f) Clive Dickson, Principal of Cairns West Primary School '93 for videoing the class
g) Brian Leighton, Maths Master of Kadina High School, Lismore Nth N.S.W '93
h) Chris Mills of Richmond River High, Lismore '93
i) Frank Broad, Ballina High '93
j) Colin Tarbox, Mullumbimby Primary Sch. '93
k) Andy Boswell '94, Gold Coast Maths Camp.
l) Ingleside Primary + Tallebudgera Qld '94
m) Stephanie Wallace, Goonellabah Primary '95
n) Gary Bell, President of the Southern Cross Mathematical Association '95
o) Jan Macks, Alstonville Primary Sch. '95
p) Susan Wynn c/o "North Coast NSW Maths & Physics Gender Equity Camp" '95.

And finally I would like to acknowledge myself for possessing the wisdom & the fortitude to publish & manifest this book.

Page 3

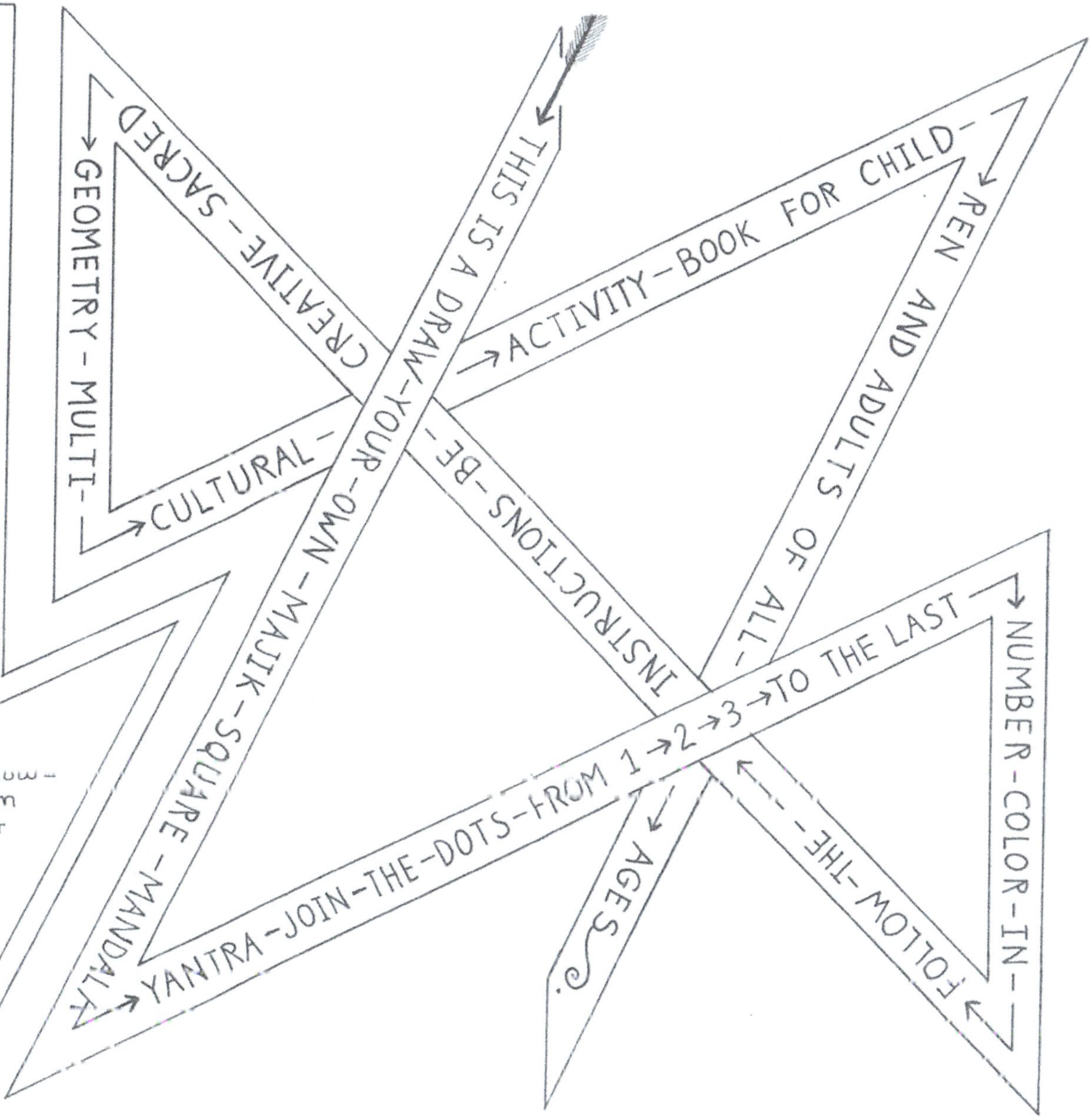

CONTENTS

THE BOOK OF HARMONY SQUARES

CONTENTS

Continued →

CONTENTS

CONTENTS

THE BOOK

CONTENTS

OF HARMONY SQUARES

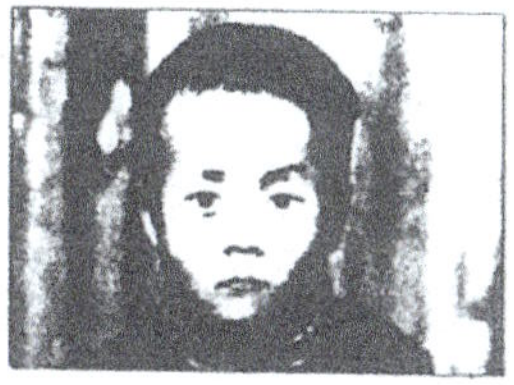

ラマ・センチェンドルジェチャン
Seṅ chen rdo rje hchaṅ
Religious Master "Seng chen Dorje chang"

Page 5

Page 6

its Time ! for Majik Sqs.

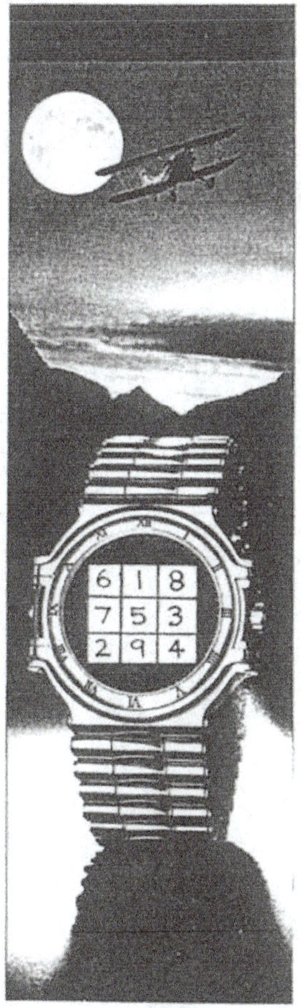

The two Tibetan astrological Calendars, shown on the following page, depicts that the Centre & Source of these Ancient Cosmological Wheels is the Lo-shu or M.Sq. of 3. The Tibetans are the "TIME-KEEPERS" of this World, keeping **RECORD** of all Incarnating Souls. To understand how M.Sqs are viewed as Frequencies or Vibrations, refer to Chapter **8** of this book on pages 23, 24, 25 & especially Page 200.

Different M.Sqs add up to different Numbers, therefore different frequencies; thus it was a Tibetan Healer or Lama's role to know which frequency would heal a person in dis-ease or dis-frequency. A Lama would write a specific M.Sq on a rice-type wafer, consecrate it & give it to the person to eat. Therefore that person was needing that frequency to heal. Today, medical computer technology has proven that all parts of the human body & its dis-eases correlate to frequencies & numbers. The Calendars below show Manjushri incarnating as a firey Tortoise who reveals the M.Sq, I-Ching & Zodiac on its Ventral. The circle of fire about the tortoise symbolizes spiritual protection. nb:- many other cultures eg China, American Indians etc have myths of a Giant Tortoise that bears the World on its back.

The 8 permutations of the M.Sq 3 are symbolized by the 8 numbers about the central cell No. 5. These formed the 8 Trigrams of the "I-Ching" (Chinese book of Oracles / Divination / Changes) which are shown as full & broken lines. (This is typical of the Computer's On-Off Binary Code / Yin-Yang etc). Then the 8 basic Trigrams form 8×8 or 64 hexagrams which is a true & ancient map of DNA. There are several books available linking the 64 Hexagrams to DNA. José Argüelles goes one step further & links the M.Sq of 8×8, whose sums are 260, with the 260 year cycles of the **Mayan** Calendar which is the most accurate calendar on this planet. Again, **Calendars**, Time, Magic Squares! Modern master geneticists have likened DNA to a 64 phenomena, that is: 4 cubed, 4×4×4 = 64.

The ancient Chinese Han Dynasty (200 BC – 200 AD) built their Palaces in 9 chambers to represent the 9 Cells of the Lo-Shu. Even their agricultural crop rotation system happened according to the M.Sq 3 eg in the 1st year, plantation was in the cell of "1", next year's plantation was in the cell of "2" which meant cell "1" was left fallow; in 3rd year, plantation was in the cell of 3 etc. This was called "MING-TANG" (which inspired the name for my first daughter – "Mingkah Sun" 'ming' is Chinese for Sun or bright, & 'Kah' is Egyptian for "Spirit"). A Prof. of M.Sqs Keith Critchlow first informed me of 'Ming Tang'.

The 9 numbers of the M.Sq3 formed an ancient Indian Calendar shown →. I have translated it from a Sanskrit copper plate given to me by an Indian Yogi. The 7 days of the week (or the 7 planets) + Sun & Moon formed a 3×3 Square. Each of the 9 M.Sqs have a different Sum or 'Constant' eg: 15, 18, 21, 24, 27 etc illustrating their different frequencies. The General M.Sq Formula for the Constants $C = \frac{1}{2} \times n \, (1 + n^2)$ is the link & secret to the onion-shelling of electrons about their nucleus. Thus, my life's work has now been validated by Atomic Physicists.

In Rajasthan, West India, a whole city of shopkeepers in Pushkar, have alcoves or recesses in their back-room walls where framed Majik Square diagrams are worshipped by lighting incense & reciting mantras. This ritual is called Puja – to invoke success into their buisness world. May a simple understandinG of these M.SQs brinG a concePt of overall HarmonY, Plan & DesiGn into Your World.

6	1	8
7	5	3
2	9	4

SUN

7	2	9
8	6	4
3	10	5

MOON

8	3	10
9	7	5
4	11	6

MARS

9	4	11
10	8	6
5	12	7

MERCURY

10	5	12
11	9	7
6	13	8

JUPITER

11	6	13
12	10	8
7	14	9

VENUS

12	7	14
13	11	9
8	15	10

SATURN

13	8	15
14	12	10
9	16	11

RAHU (Nth Node of ☽)

14	9	16
15	13	11
10	17	12

KETU (Sth Node of ☽)

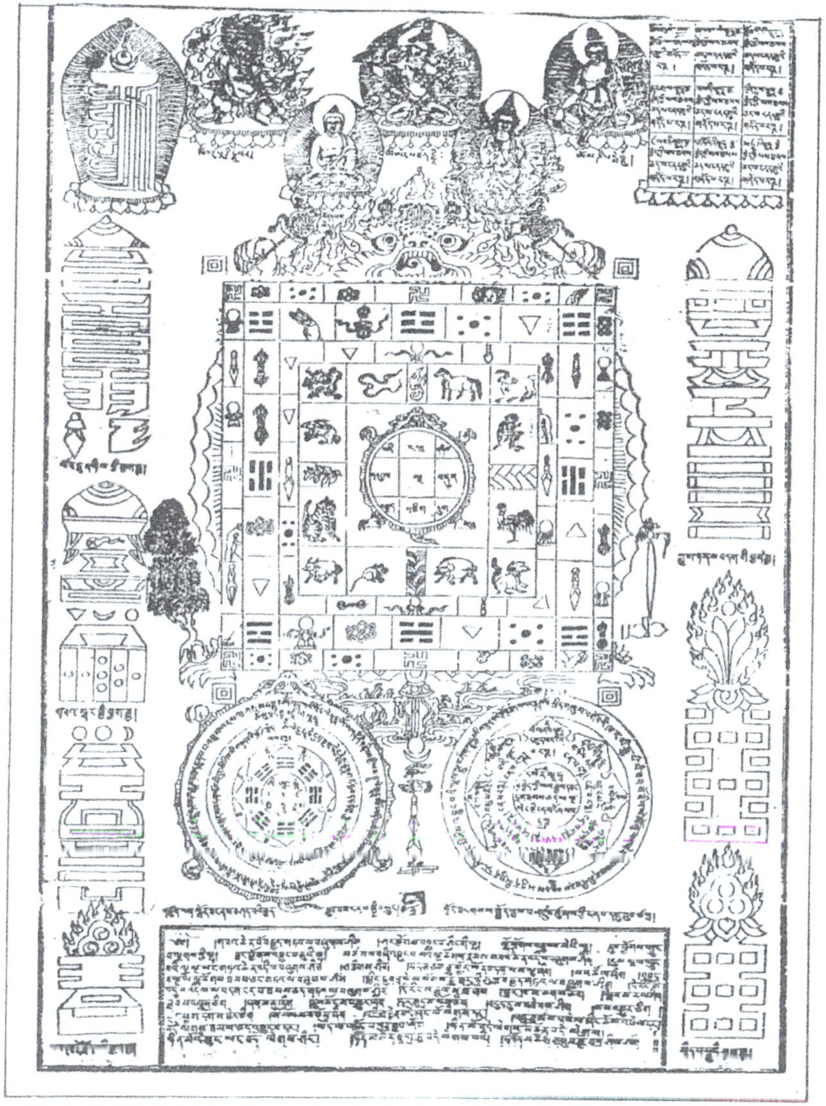

Page 7

THE CENTRE OF THE TIBETAN CALENDAR IS THE M. SQ. OF 3×3 !

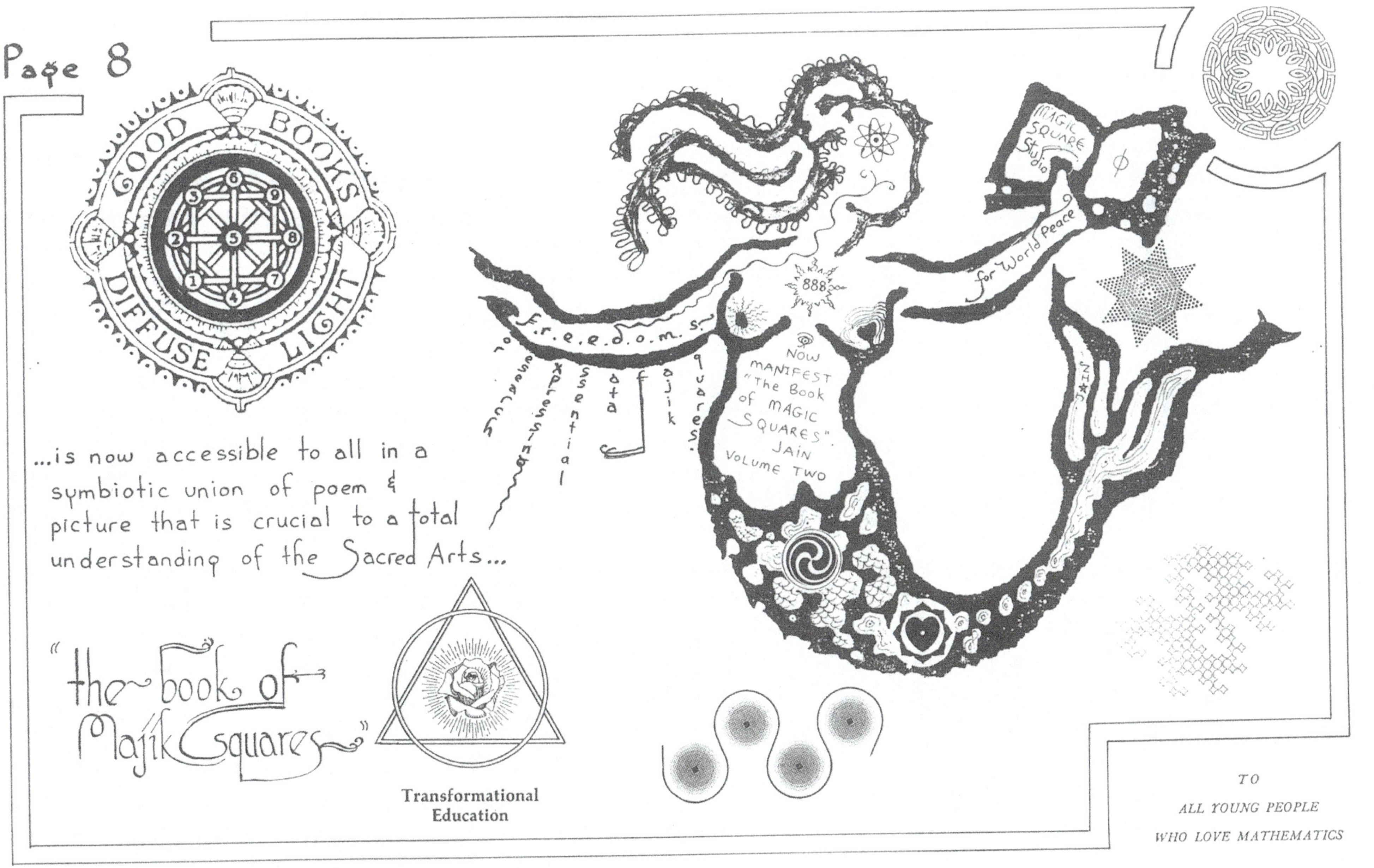

Page 8
GOOD BOOKS DIFFUSE LIGHT
...is now accessible to all in a symbiotic union of poem & picture that is crucial to a total understanding of the Sacred Arts...
"the book of Majik Squares"
Transformational Education
MAGIC SQUARE Studio
for World Peace
f.r.e.e.d.o.m.s
888
NOW MANIFEST "The Book of MAGIC SQUARES". JAIN VOLUME TWO
TO
ALL YOUNG PEOPLE
WHO LOVE MATHEMATICS

A briefe and true **guide to the** كتاب المربعات السحرية

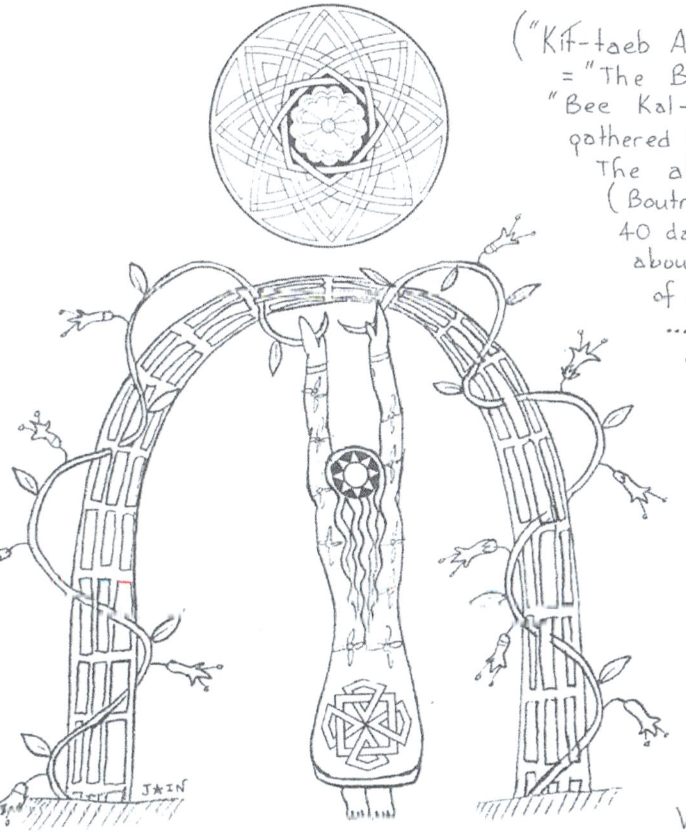

("Kit-taeb Al-Moo-rub-ba-ought El-see-rhee-yah = "The Book of Majik Squares" "Bee Kal-lum Jain" = "By Jain, as gathered by Jain, in service").

بقلم جاين

The above titles, in Lebanese script were written by my Earth Father (Boutros Saad) who visited me with his wife Yvonne, just as I had broke a 40 day Wheatgrass Juice + Shivambu fast. For 3 days we talked things over, about Unconditional Love, Non-Attachment, Compassion, sharing & distribution of wealth (they're multi-millionaires!), Majik Squares & Golden Proportions ... nephews & neices, New Age Printeries on The Farms, buying this olde dilapidated school house... Cups of teas & cookies... & we talked about world travels, marriage to Arabian virgins, wheatgrass & buckwheat sprouts, Healing by being & feeling vital, tabouli & egg-plant gardens, FlowerLands... We decided to go to the mountain-top & splished & splashed & bubbled in an immense river at the head of the Ellen-borough Falls which is the largest single drop of water in the Southern Hemisphere (or so they say)... And Dad looked great immersed & rejuvenated as our Eyes melted & oozed rays of Love. We layed in a mighty massage pool, our Auraes knitted closely, & like 2 little boys we watched Mum cupping the cold water over her large breasts... this ReUnion was their 1st visit 2 me in 10 years since I scurried from Home when I disobeyed my Dad's orders to study medicine & I did for the first time what I wanted to do: Art & Writing. It meant so much to me, this Moment Now, this Family Love, that this Book of Doodles, this Book of Majik Squares has become a rare & sacred Vessel 4 me 2 pour my Heart in 2. Jain, Nov '86 Kippaxs school house.

Page 9

Page 10

Our deepest fear is not that we are inadequate. Our deepest fear is that we are powerful beyond measure. It is our light, not our darkness, that most frightens us.

We ask ourselves, who am I to be brilliant, gorgeous, - talented and fabulous? Actually, who are you NOT to be? You are a child of God. Your playing small doesn't serve the world.

There's nothing enlightened about shrinking so that other people won't feel insecure about you.

We were born to make manifest the glory of God that is within us. It's not just in some of us; it's in everyone. And as we let our own light shine, we unconsciously give other people permission to do the same.

As we are liberated from our own fear, our presence automatically liberates others.

1994 Inaugural Speech
Nelson Mandela

eight

BOOK
OF
NUMBERS

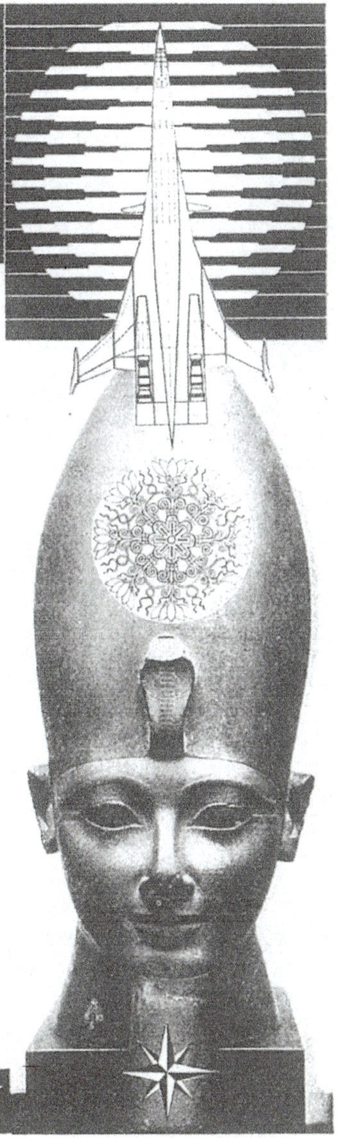

MAJIK SQUARES OF EIGHT

EDUCATION BY NUMBERS

HARMONIE UND PRÄZISION

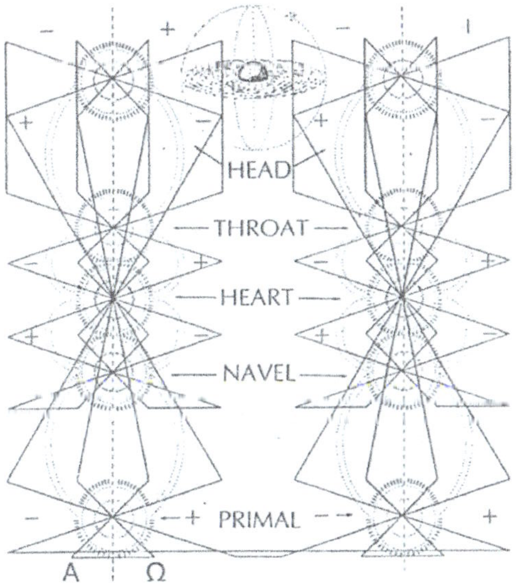

INTRODUCING THE BENJAMIN FRANKLIN MAJIK SQUARE OF EIGHT

Magic Squares of Eight and Sixteen, Benjamin Franklin's Contribution

260	260	260	260	260	260	260	260	
52	61	4	13	20	29	36	45	260
14	3	62	51	46	35	30	19	260
53	60	5	12	21	28	37	44	260
11	6	59	54	43	38	27	22	260
55	58	7	10	23	26	39	42	260
9	8	57	56	41	40	25	24	260
50	63	2	15	18	31	34	47	260
16	1	64	49	48	33	32	17	260

ORIGINAL FRANKLIN SQUARE, ARRAY OF 8

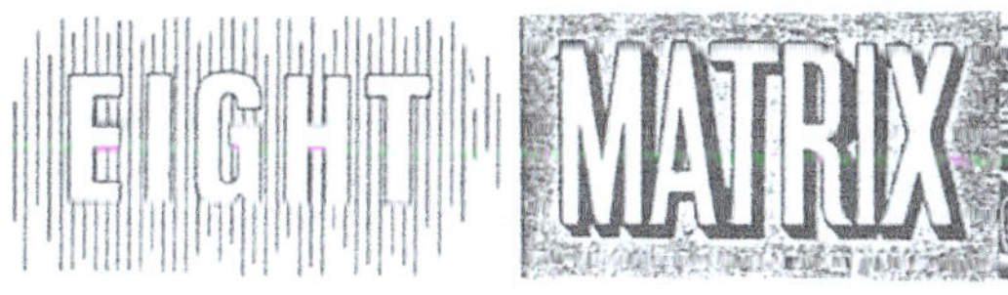

Page 11

An Analysis of the Symmetrical Features of the Franklin Square of Eight

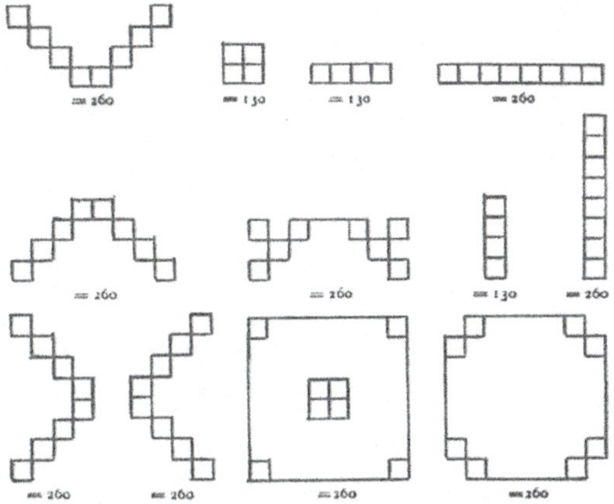

Fig 2

- The 8 Columns & The 8 Rows have sums or Constants of 260 as shown graphically in Fig 2 above.
- Any 4 cells in Vertical or Horizontal array sum 130.
- Any 4 cells in 2×2 sub-squares = 130.
- The bent diagonals of 4 cells up & 4 cells down sum 260
- The 4 corner cells plus the 4 central cells sum 260.
- The sum of any 4 cells that lie equidistant from the Square's centre is 260.

FIG 1

Format For The Arrangement Of The Following Pages.

This book is designed for children & adults as a creative Coloring - In Book where all the designs are created by the reader by drawing a long continuous line from beginning to end of a box of numbers, known as a Majik Square or Harmony Square. All Columns, Rows & Diagonals have a common sum, known as The CONSTANT which is the secret coding for the ancient numerical names of Allah (God). When the last number ("Ω" Omega) is joined to the 1st number 1 ("α" Alpha) an Eternal bio-circuit of Life Force Prana is in flow, often replicating the Pyramid's Proportions.

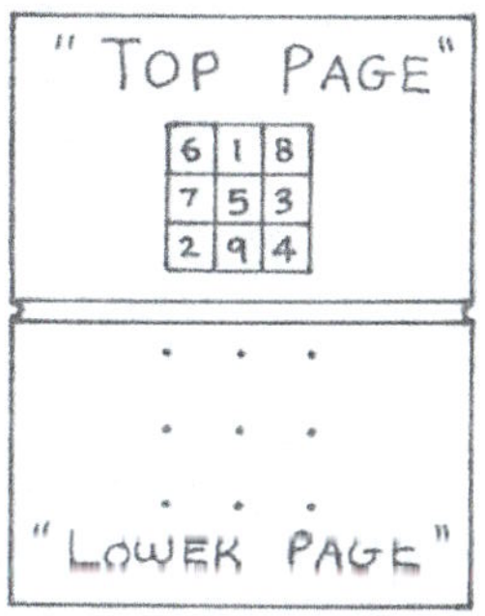

To assist in easy methods of construction, as learnt from the difficulties that arose from the former book "Volume One", the Magic Squares will appear on the "TOP" pages of this book so that the numbers can be quickly read. Then the Magic Square Pattern / Diagram / Yantra can be drawn on the "LOWER" page of this open book so that the construction is close to the Heart. This makes it easy to recognise that whenever you see a Magic Square, there will always be a corresponding grid or lattice of Dots nearby. Enjoi.

follow not any Monasticism, ie one belief like Christians who believe only in X⁺, or Buddhists who believe only in Buddha, but rather, let your Heart be winged; Embrace all Faiths of the Light of the Ages. Life's a Whirl; this is the Way of The Whirling dervish.

Be Light
Be Happy

Be Love
Be Only
who U R
Be free &
Play all day.

Page 13

The Benjamin Franklin Magik Square of 8×8.

52	61	4	13	20	29	36	45
14	3	62	51	46	35	30	19
53	60	5	12	21	28	37	44
11	6	59	54	43	38	27	22
55	58	7	10	23	26	39	42
9	8	57	56	41	40	25	24
50	63	2	15	18	31	34	47
16	1	64	49	48	33	32	17

הראבד :
הנתיב הלא נקר שבל תמידי
כי הוא המנהיג השמש והירח
ושאר הכוכבים והצורות כל
אחד מהם בגלגלו ונותן לכל
הנבראים ממערכתם אל
המזלות והצורות:

Fig 1

In the Space Provided of Fig 3 (Grid Matrix of 64 dots) join a long continuous line, by referring to Fig 1 above, from Number 1 → 2 → 3 → 4 etc to → 64 (the last Number). Then draw a final Line from 64 → 1 to close the Circuit. This diagram is called a "YANTRA".

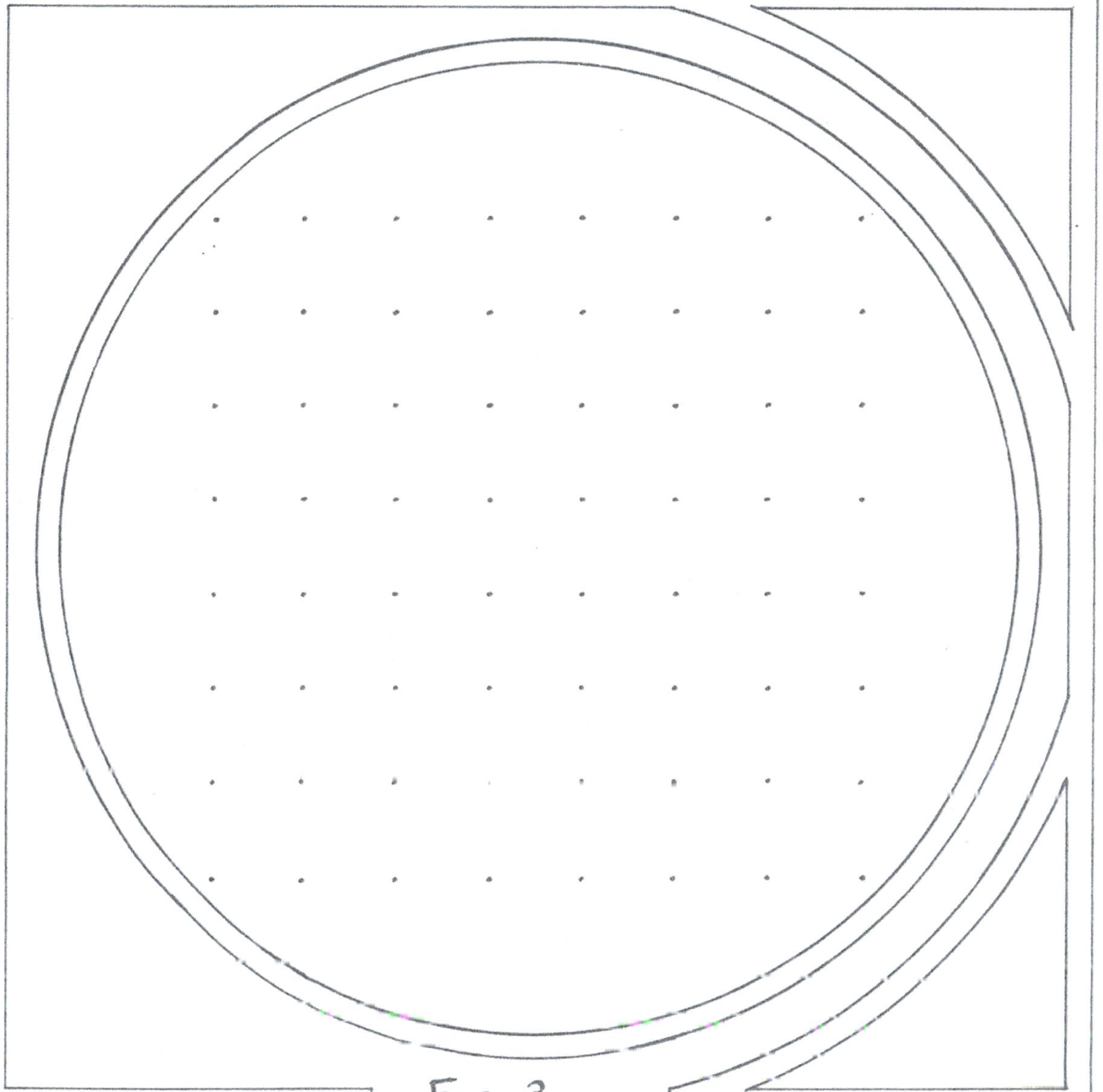

FIG. 3.

Page 15

Page 16

FIG. 4

52	61	4	13	20	29	36	45
14	3	62	51	46	35	30	19
53	60	5	12	21	28	37	44
11	6	59	54	43	38	27	22
55	58	7	10	23	26	39	42
9	8	57	56	41	40	25	24
50	63	2	15	18	31	34	47
16	1	64	49	48	33	32	17

Shown here are the same 4 Franklin Harmony Squares of 8 that will be necessary for the construction of Fig 8 (Rotated at 0° + 90° + 180° + 270°).

In the space of Fig 8, Fig 4 will be drawn first. Because it is already upright, according to our seated position, it is expressed as 0°.

Then Fig 5 is superimposed at a 90° Rotation, in the space of Fig 8. So the book will have to be turned around until the numbers are upright.

Fig 6 is at 180°. Turn the book around 180° & superimpose this upon Figs 4 & 5, in the space provided of Fig 8.

Fig 7 is at 270°. Similarly, turn the book around until Fig 7 is upright & superimpose this upon the combined Yantras in the space of Fig 8.

FIG. 5

16	50	9	55	11	53	14	52
1	63	8	58	6	60	3	61
64	2	57	7	59	5	62	4
49	15	56	10	54	12	51	13
48	18	41	23	43	21	46	20
33	31	40	26	38	28	35	29
32	34	25	39	27	37	30	36
17	47	24	42	22	44	19	45

FIG. 6

17	32	33	48	49	64	1	16
47	34	31	18	15	2	63	50
24	25	40	41	56	57	8	9
42	39	26	23	10	7	58	55
22	27	38	43	54	59	6	11
44	37	28	21	12	5	60	53
19	30	35	46	51	62	3	14
45	36	29	20	13	4	61	52

FIG. 7

45	19	44	22	42	24	47	17
36	30	37	27	39	25	34	32
29	35	28	38	26	40	31	33
20	46	21	43	23	41	18	48
13	51	12	54	10	56	15	49
4	62	5	59	7	57	2	64
61	3	60	6	58	8	63	1
52	14	53	11	55	9	50	16

Fig 8

The lattice of 64 dots on your right is Fig 8 upon which you will construct The B. Franklin Harmony Square of 8 at 4 × 90° (ie: 0° + 90° + 180° + 270°). Refer to "The Solutions" on p 198.

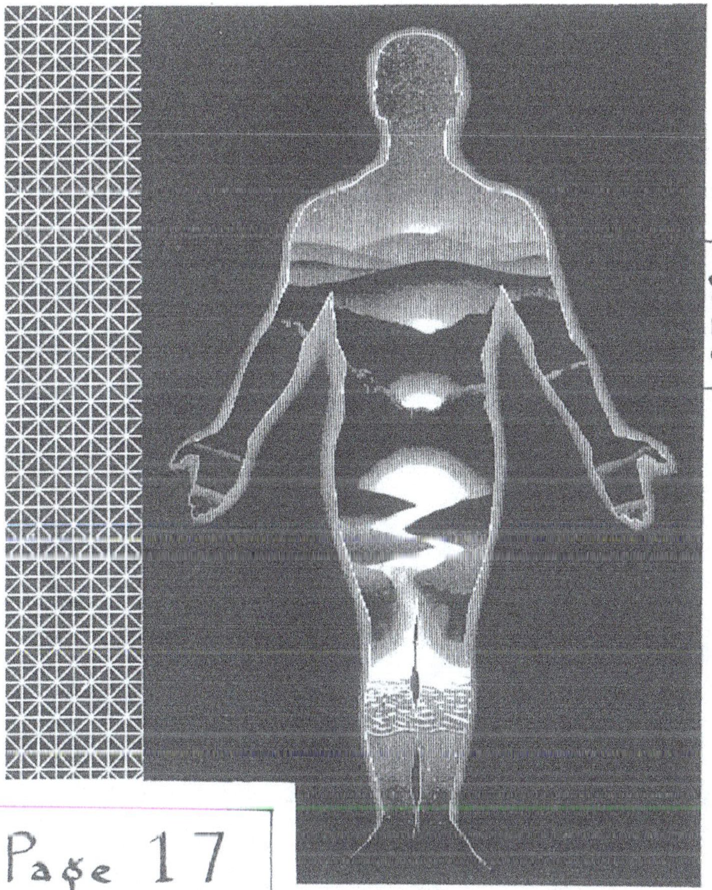

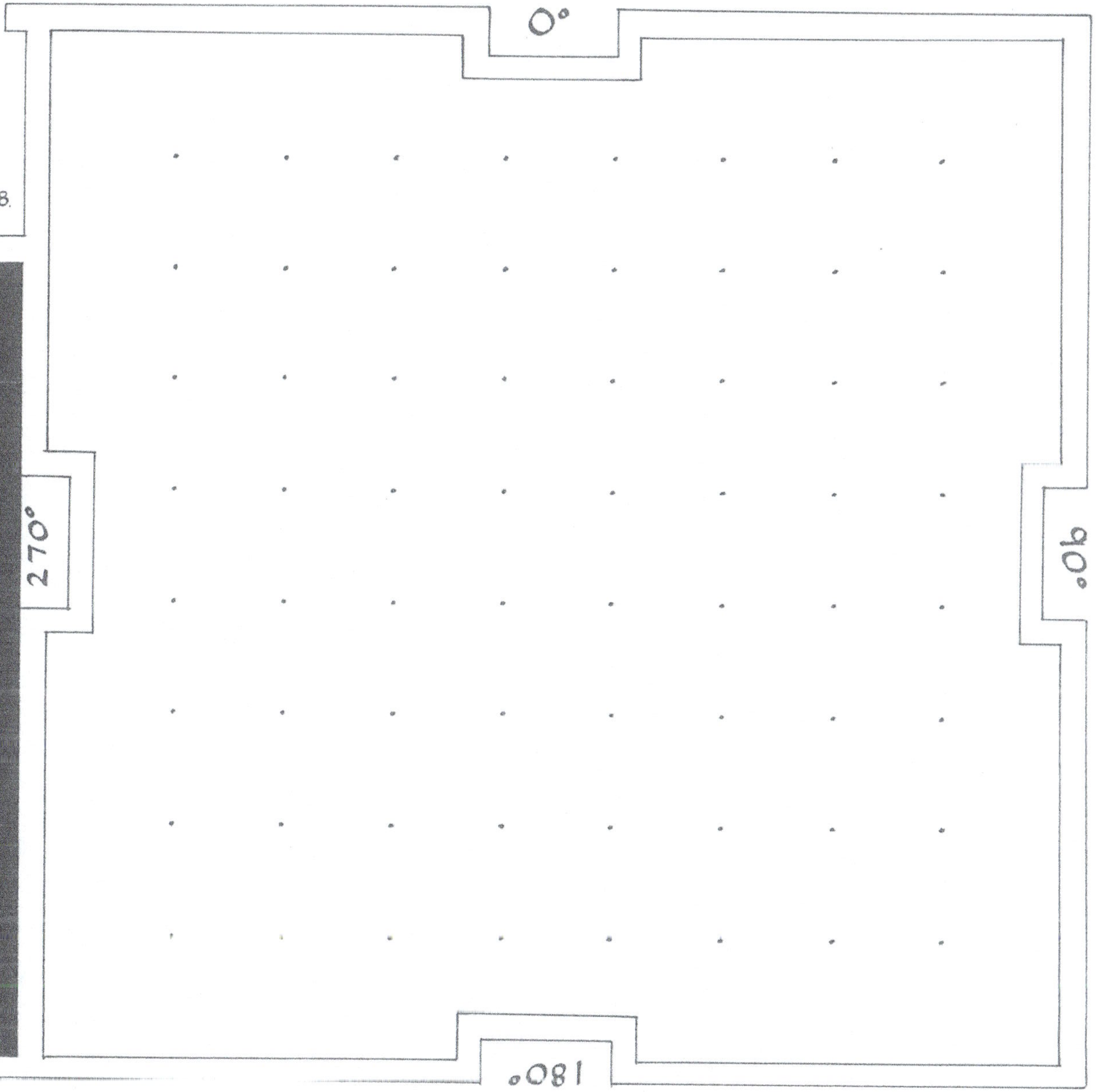

Page 17

ISOLATING AND HIGHLIGHTING HARMONY SQ. PATTERNS.

Fig. 9 is a completed Franklin Harmony Sq. of 8 which you will not be required to draw. I will outline in 4 steps how I arrived at this pattern. This is to express to you that you are unlimited in your approach to forming original & creative patterns. Each individual may perceive Harmony Square Art in totally different ways & this is to be encouraged.

Step 1 : By referring to Fig 1 on p 19 connect only The ODD numbers. Let us call this pattern Diagram "A".

Step 2 : On another Lattice of 64 Dots, connect all The EVEN numbers. Call this pattern Diagram "B". You will notice that Diagram "A" is the Mirror Image of Diagram "B" This is an important Realization.

Step 3 : Starting again, draw "B" super-imposed upon "A" at 0°, i.e Draw a combined pattern of The Odds over The Evens. Call this Diagram "C".

Step 4 : Diagram "C" is then drawn four times i.e at 0° + 90° + 180° +270°. This results in Fig 9. By Meditating upon this Yantra, each individual will color-in a certain facet of this Heart Crystal.

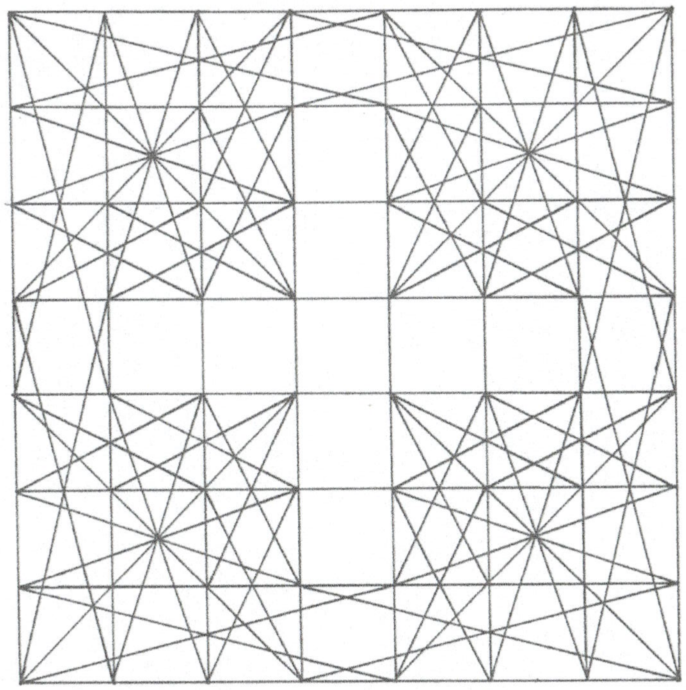

Fig 9.

Page 18

Fig 1

52	61	4	13	20	29	36	45
14	3	62	51	46	35	30	19
53	60	5	12	21	28	37	44
11	6	59	54	43	38	27	22
55	58	7	10	23	26	39	42
9	8	57	56	41	40	25	24
50	63	2	15	18	31	34	47
16	1	64	49	48	33	32	17

In arithmetic we create a universe of figures by the process of counting; in geometry we create another universe by drawing lines in the abstract field of imagination, laying down definite directions; in algebra we produce magnitudes of a still more abstract nature, expressed by letters. In all these cases the first step producing the general conditions in which we move, lays down the rule to which all further steps are subject, and so every one of these universes is dominated by a consistency, producing a wonderful symmetry.

There is no science that teaches the harmonies of nature more clearly than mathematics, and the magic squares are like a mirror which reflects the symmetry of the divine norm immanent in all things, in the immeasurable immensity of the cosmos and in the construction of the atom not less than in the mysterious depths of the human mind.

Paul Carus.

Fig 10

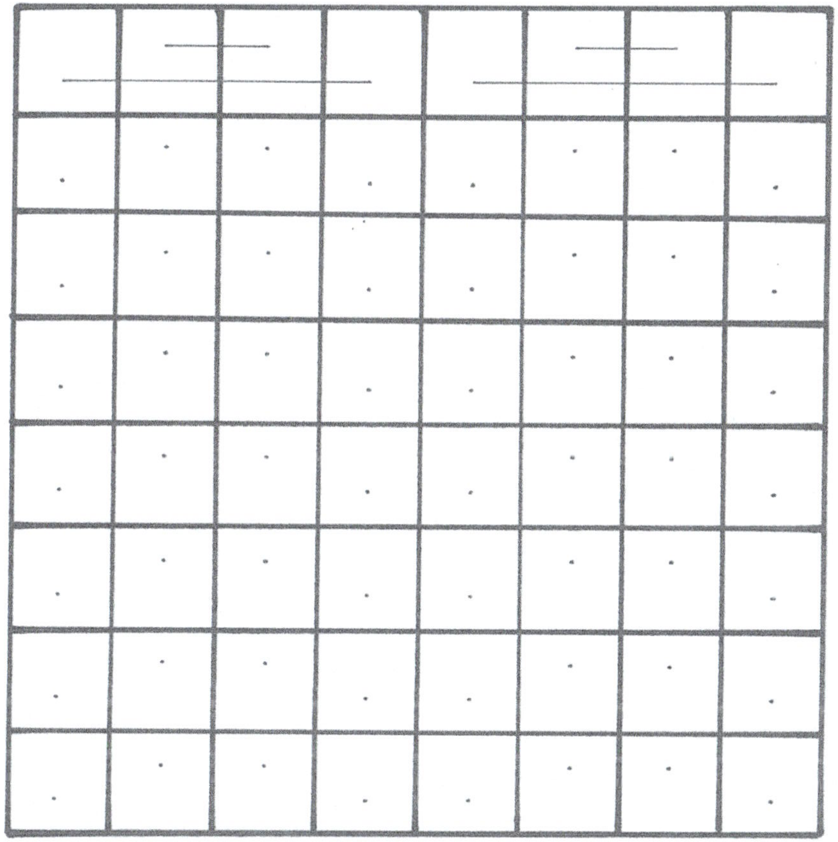

The Balance of the Pairs

of Benjamin Franklin's Harmony Square of Eight.

By studying carefully Fig 1 observe all the pairs of numbers that have a sum of 65. This number is determined by a certain formula $(1 + n^2)$ where 'n' is the size or magnitude of the harmony square. Here, $n = 8$. Thus the first Number (1) + The last Number (64 or n^2) = **65**. In the space of Fig 10, link or draw a line to each pair that sums 65. The top line of Fig 10 is already completed. Use the dots provided to illustrate the Internal Harmony!

COLOR CODING of MAJIK SQs

The diagram on this page, known as Fig 11, which you will complete, is known as "THE B. FRANKLIN HARMONY SQUARE MOD 8". Any Harmony Square, like Fig 1, which you are already familiar with, can be colored in according to a special coding. Fig 1 contains single digits and double digits. Double digited numbers e.g: 17, 33, 64 etc can be converted to single digits. We will convert the 64 digits of Fig 2 into 8 different single digits: "1, 2, 3, 4, 5, 6, 7, 0". This changes the whole of the Harmony Square to "Mod 8". We begin by writing, in the space of Fig 11, the numbers 1 to 7 of the Harmony Square in the same place as they already are. When we get to the No. 8 we write "0" and then the cycle of 8 digits begins again. Thus 9 becomes 1, 10 → 2, 11 → 3, 12 → 4, 13 → 5, 14 → 6, 15 → 7, 16 → 0. Notice that we replace every multiple of 8 with "0". (The multiples of 8 are: 8, 16, 24, 32, 40, 48, 56 & 64).

To assist you, the first 8 digits of Fig 11 have been entered in. Complete the square until all 64 cells have been written in Mod 8.

When any activity in this book has been completed you can check to see if your work is correct by referring to the end chapter known as "CHAPTER ZERO: SOLUTIONS TO THE HARMONY SQUARES".

52	61	4	13	20	29	36	45
14	3	62	51	46	35	30	19
53	60	5	12	21	28	37	44
11	6	59	54	43	38	27	22
55	58	7	10	23	26	39	42
9	8	57	56	41	40	25	24
50	63	2	15	18	31	34	47
16	1	64	49	48	33	32	17

FIG 1.

		4					
	3						
		5					
	6						
		7					
	0						
		2					
	1						

FIG 11.

The box of 64 squares, Fig 12 on the next page is the space to select & use 8 different Colors of your choice.

COLOUR CODING OF HARMONY SQUARES, Continued.

THE method of Color Coding Harmony Squares is very simple. Just select 8 different Colors of your choice. All the "0"s, of Fig 11 will be one color; all the "1"s will be another color; all the "2"s another; etc until the whole box of Fig 12 is completely colored-in. As you Color-In, observe the symmetrical arrangement of the 8 Colors. Use this concept in your house designs, like Tiles.

Fig 12

THIS EARTH IS PRECIOUS

How can you buy or sell the sky, the warmth of the land? The idea is strange to us. If we do not own the freshness of the air and the sparkle of the water, how can you buy them?

Every part of this earth is sacred to my people. Every shining pine needle, every sandy shore, every mist in the dark woods, every clearing and humming insect is holy in the memory and experience of my people. The sap which courses through the trees carries the memories of the red man. The white man's dead forget the country of their birth when they go to walk among the stars. Our dead never forget this beautiful earth, for it is the mother of the red man. We are part of the earth and it is part of us. The perfumed flowers are our sisters; the deer, the horse, the great eagle, these are our brothers. The rocky crests, the juices in the meadows, the body heat of the pony, and man -- all belong to the same family.

So, when the Great Chief in Washington sends word that he wishes to buy our land, he asks much of us. The Great Chief sends word he will reserve us a place so that we can live comfortably to ourselves. He will be our father and we will be his children. So we will consider your offer to buy our land. But it will not be easy. For this land is sacred to us. This shining water that moves the streams and rivers is not just water but the blood of our ancestors. If we sell you land, you must remember that it is sacred, and you must teach your children that it is sacred and that each ghostly reflection in the clear water of the lakes tells of events and memories in the life of my people. The water's murmur is the voice of my father's father.

The rivers are our brothers, they quench our thirst. The rivers carry our canoes, and feed our children. If we sell you our land, you must remember, and teach your children, that the rivers are our brothers, and yours, and you must henceforth give the rivers the kindness you would give any brother. We know that the white man does not understand our ways. One portion of land is the same to him as the next for he is a stranger who comes in the night and takes from the land whatever he needs. The earth is not his brother, but his enemy, and when he has conquered it, he moves on. He leaves his father's graves behind, and he does not care. He kidnaps the earth from his children, and he does not care. His father's grave, and his children's birthright, are forgotten. He treats his mother, the earth, and his brother, the sky, as things to be bought, plundered, sold like sheep or bright beads. His appetite will devour the earth and leave behind only a desert. I do not know. Our ways are different from your ways. The sight of your cities pains the eyes of the red man. But perhaps it is because the red man is a savage and does not understand. There is not quiet place in the white man's cities. No place to hear the unfurling of leaves in spring, or the rustle of the insect's wings. But perhaps it is because I am a savage and do not understand. The clatter only seems to insult the ears. And what is there to life if a man cannot hear the lonely cry of the whippoorwill or the arguments of the frogs around a pond at night? I am a red man and do not understand. The Indian prefers the soft sound of the wind darting over the face of a pond, and the smell of the wind itself, cleaned by a midday rain, or scented with the pinon pine.

The air is precious to the red man, for all things share the same breath -- the beast, the tree, the man, they all share the same breath. The white man does not seem to notice the air he breathes. Like a man dying for many days, he is numb to the stench. But if we well you our land, you must remember that the air is precious to us, that the air shares its spirit with all the life it supports. The wind that gave our grandfather his first breath also receives his last sight. And if we well you our land, you must keep it apart and sacred, as a place where event the white man can go to taste the wind that is sweetened by the meadow's flowers.

So we will consider your offer to buy our land. If we decide to accept, I will make one condition: The white man must treat the beasts of this land as his brothers. I am a savage and I do not understand any other way. I have seen a thousand rotting buffaloes on the prairie, left by the white man who shot them from a passing train. I am a savage and I do not understand how the smoking iron horse can be more important than the buffaloe that we kill only to stay alive. What is man without the beasts? If all the beasts were gone, man would die from a great loneliness of spirit. For whatever happens to the beasts, soon happens to man. All things are connected.

You must teach your children that the ground beneath their feet is the ashes of your grandfathers. So that they will respect the land, tell your children that the earth is rich with the lives of our kin. Teach your children what we have taught our children, that the earth is our mother. Whatever befalls the earth befalls the sons of the earth. If men spit upon the ground, they spit upon themselves. This we know: The earth does not belong to man; man belongs to the earth. This we know. All things are connected like the brook which unites one family. all things are connected. Whatever befalls the earth befalls the sons of the earth. Man did not weave the web of life: he is merely a strand in it. Whatever he does to the web, he does to himself. Even the white man, whose God walks and talks with him as friend to friend, cannot be exempt from the common destiny. We may be brothers after all. We shall see. One thing we know, which the white man may one day discover -- our God is the same God. You may think now that you own Him as you wish to own our land; but you cannot. He is the God of man, and His compassion is equal for the red man and the white. This earth is precious to Him, and to harm the earth is to heap contempt on its Creator. The whites too shall pass; perhaps sooner than all other tribes. Contaminate your bed, and you will one night suffocate in your own waste. But in your perishing you will shine brightly, fired by the strength of the God who brought you to this land and for some special purpose gave you dominion over this land and over the red man. That destiny is a mystery to us, for we do not understand when the buffalo are all slaughtered, the wild horses are tames, the secret corners of the forest heavy with scent of many men, and the view of the ripe hills blotted by talking wires. Where is the thicket? Gone. Where is the eagle? Gone. The end of living and the beginning of survival.

CHIEF SEATTLE, DUWAMISH TRIBE

HOW TO CONVERT MAJIK SQUARES INTO WAVE PATTERNS

MAJIK SQUARES: AN ANALYSIS OF CONSECUTIVE NUMBER SEQUENCE BY ROWS.

In June '92, I met Prof. Callum Coats who researches Patterns in Nature. After giving him my Vol 1 m.Sq book, he returned home. Months later he sent me the information known here as Fig 14 which shows the wave patterns of 6 Magick Sqs, analysing each square row by row.
e.g: shown below is the famous Lo-shu (m.Sq of 3). Its top row depicts the 3 digits "6, 1, 8" We now plot those digits wherever they appear in a natural line of 1 to 9.

6	1	8
7	5	3
2	9	4

(1) 2 3 4 5 (6) 7 (8) 9 ←

Then we observe the 2nd row "7, 5, 3" and circle those digits.

1 2 (3) 4 (5) 6 (7) 8 9 ←

And then the last row "2, 9, 4".

1 (2) 3 (4) 5 6 7 8 (9) ←

Fig 13 shows the same pattern in dot form & with a line that joins all the 9 Nos. in their consecutive order.

FIG 13

Page 24

ANALYSIS OF CONSECUTIVE NUMBER SEQUENCE BY ROWS

Prof. Callum Coats

June '92

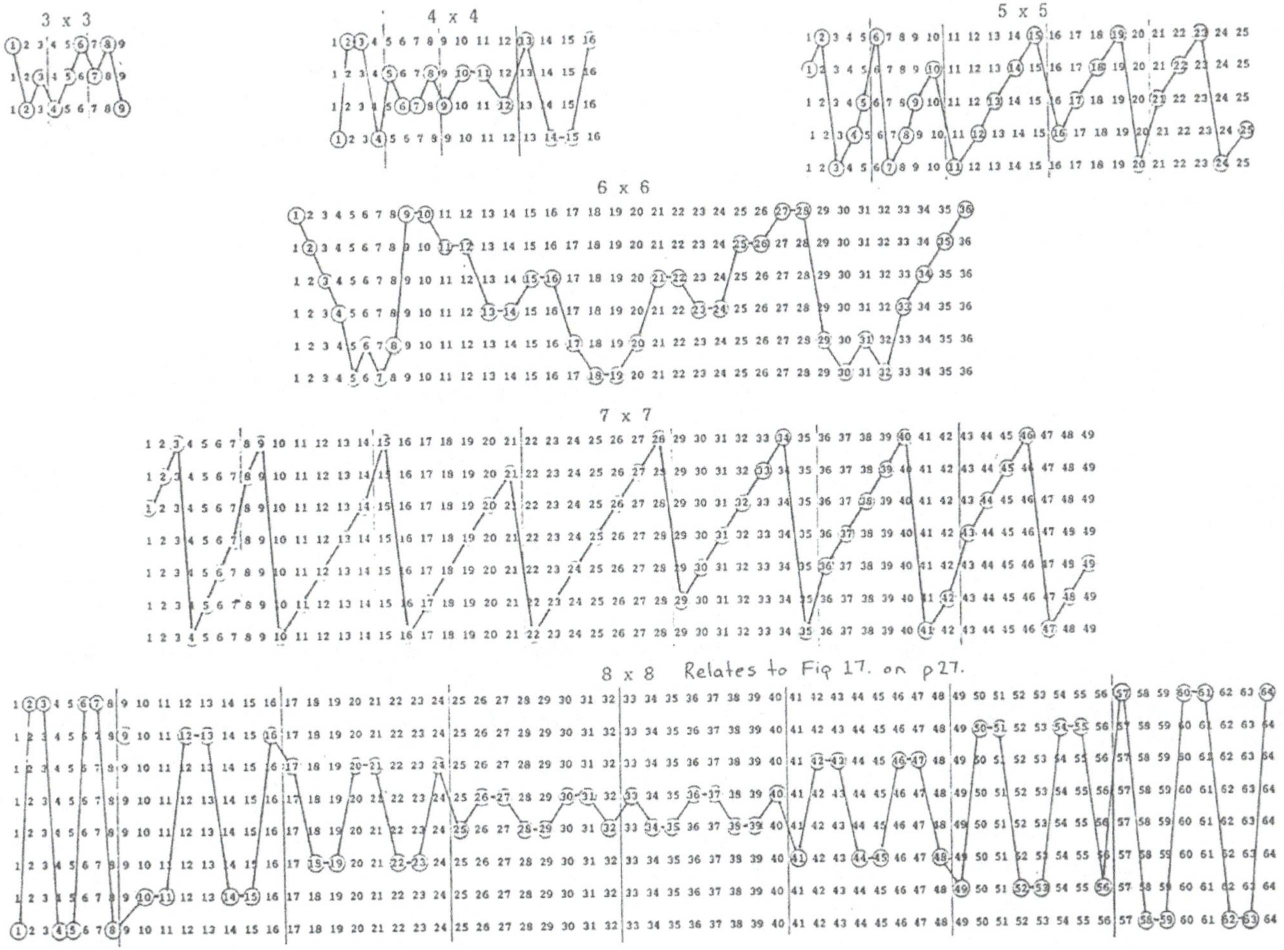

FIG 14

An ANALYSIS OF CONSECUTIVE NUMBER SEQUENCE BY ROWS.

FIG 15

On the lower right-hand corner is the Benjamin Franklin Harmony Sq. of 8×8. The Grid of 8×64 dots above, known as Fig 15 is the space provided for you to plot all the Nos. of each row of the Harmony (or Majik) Square in its appropriate position. To get you started, the top row has been plotted in by using a heavier dot. When all the rows have been plotted, begin to join the wave pattern by proceeding from 1→2→3→4 etc →64. (Where the peaks & troughs occur you may choose to use sharp points or curve them freehand).

"THE WORLD IS BUILT UPON THE POWER OF NUMBERS." quote from: PYTHAGORAS C 6th BC.

52	61	4	13	20	29	36	45
14	3	62	51	46	35	30	19
53	60	5	12	21	28	37	44
11	6	59	54	43	38	27	22
55	58	7	10	23	26	39	42
9	8	57	56	41	40	25	24
50	63	2	15	18	31	34	47
16	1	64	49	48	33	32	17

Page 25

"If I can't picture it, I can't understand it". Albert Einstein.

INTRODUCING OTHER MAJIK SQUARES OF

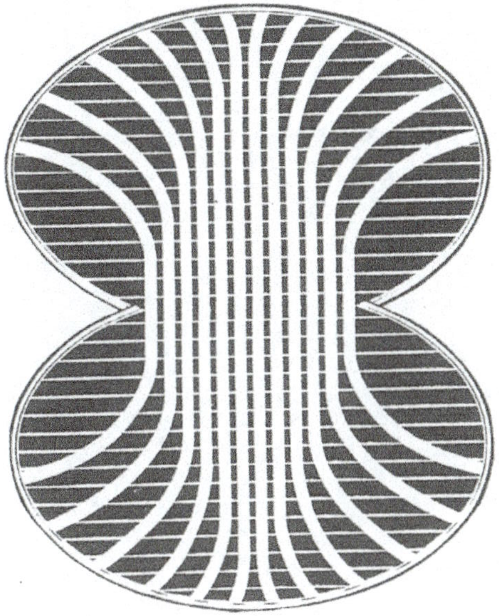

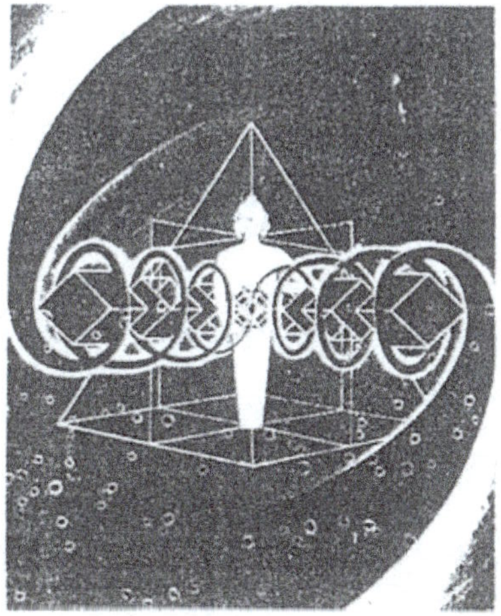

CONSTRUCTION OF A HARMONY SQUARE OF 8×8.

Begin with a NATURAL SQUARE of 8 as shown in Fig 16: (a box of the natural counting order from $1 \rightarrow 64$). View the whole 8×8 in its quadrants. Each of the 4 quadrants would have 16 cells i.e: 4×4 sub-squares. Observe where the main diagonals are of each quadrant of Fig 16. Overall, the main diagonals of Fig 16 are shown as the dotted lines. Notice that the sum of the two main diagonals of this Natural Square both have a sum of 260 which equals the sum of the 4 sub-diagonals = 520 !!!

To form a Harmony Square of 8, each cell or number of all the diagonals will be replaced by its COMPLEMENT. The Complement of any No. is the No. that you must add to it to make the sum of "The first & the last No. of the Harmony Sq." (i.e: $1+n^2$, where n = the magnitude of the Harmony Sq. = 8 Thus $1+n^2 = 1+64 = 65$. When all the Nos. of the diagonals have been replaced by their complement, the remaining Nos. of Fig 16 are inserted, unchanged, into the same diagram. This is shown in Fig. 16a which transforms the Natural Square of 8 into a Harmony Sq. of 8 where all Columns, Rows & Diagonals sum 260. This sum of equality is known as the 'Constant'. This Constant (C = 260) is true for all Harmony Sqs of 8. It exists in the diagonals of all Natural Squares, not just this 8x8. For construction purposes, you could say that the Diagonals are Reversed. This method of Construction is true for Harmony Sqs. of 4, 8, 12, 16, 20 etc. i.e: for all Multiples of 4.

1	2	3	4	5	6	7	8
9	10	11	12	13	14	15	16
17	18	19	20	21	22	23	24
25	26	27	28	29	30	31	32
33	34	35	36	37	38	39	40
41	42	43	44	45	46	47	48
49	50	51	52	53	54	55	56
57	58	59	60	61	62	63	64

Fig 16

	2	3			6	7	
9			12	13			16
17			20	21			24
	26	27			30	31	
	34	35			38	39	
41			44	45			48
49			52	53			56
	58	59			62	63	

64	2	3	61	60	6	7	57
9	55	54	12	13	51	50	16
17	47	46	20	21	43	42	24
40	26	27	37	36	30	31	33
32	34	35	29	28	38	39	25
41	23	22	44	45	19	18	48
49	15	14	52	53	11	10	56
8	58	59	5	4	62	63	1

Fig 16a

64	2	3	61	60	6	7	57
9	55	54	12	13	51	50	16
17	47	46	20	21	43	42	24
40	26	27	37	36	30	31	33
32	34	35	29	28	38	39	25
41	23	22	44	45	19	18	48
49	15	14	52	53	11	10	56
8	58	59	5	4	62	63	1

Fig 17

Fig 18 shown here is the same Harmony Square of 8 that was constructed in Figs 16 & 17. The only difference is that the top rows of Fig 17 are interchanged with the bottom rows. Regardless of this 180° reflection of the numbers, both Harmony Sqs. produce the same graphic / yantra.

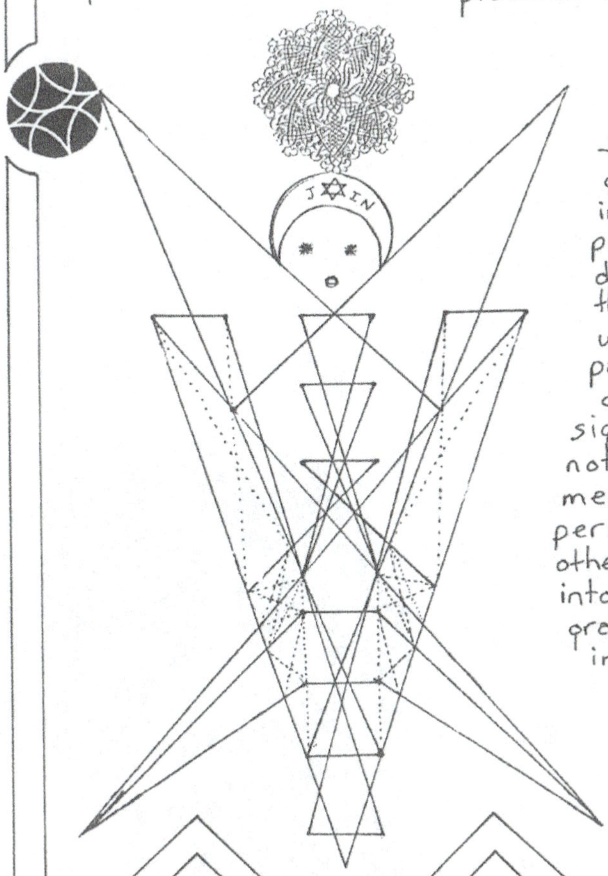

The diagram on your left is an example of how you can take segments of Harmony Square Art & use these geometric creations to inspire your doodles. This particular segment is derived from Fig 18 the whole of which you will draw on the following page.

Any ...dles ... book signed ... not in ... means ... in... personal ... ror. other arti... work into this ...pend graphics ... ackno in the back of th...

Fig 18

8	58	59	5	4	62	63	1
49	15	14	52	53	11	10	56
41	23	22	44	45	19	18	48
32	34	35	29	28	38	39	25
40	26	27	37	36	30	31	33
17	47	46	20	21	43	42	24
9	55	54	12	13	51	50	16
64	2	3	61	60	6	7	57

LIMITED EDITION

new art

Fig 19

In the space of Fig 19 above draw the Majik Square Yantra based on Fig 18.

Page 29

Some ancient Qabalahistic texts ascribe the 8×8 Majik Square as the Archetype of → MERCURY

MERCURY

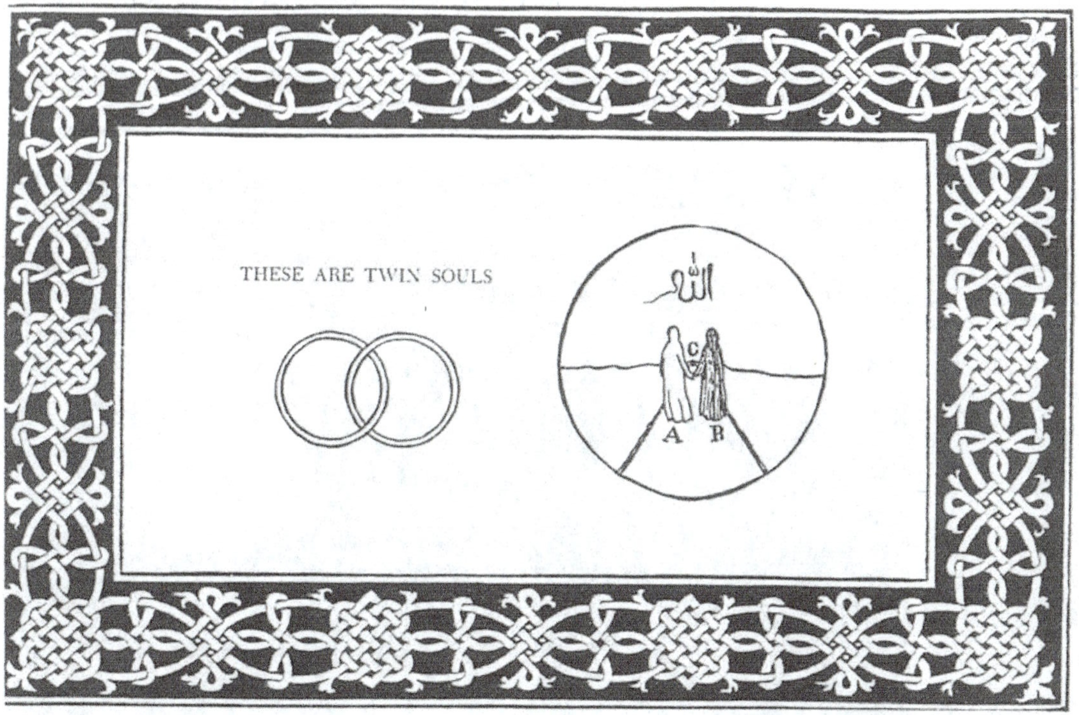

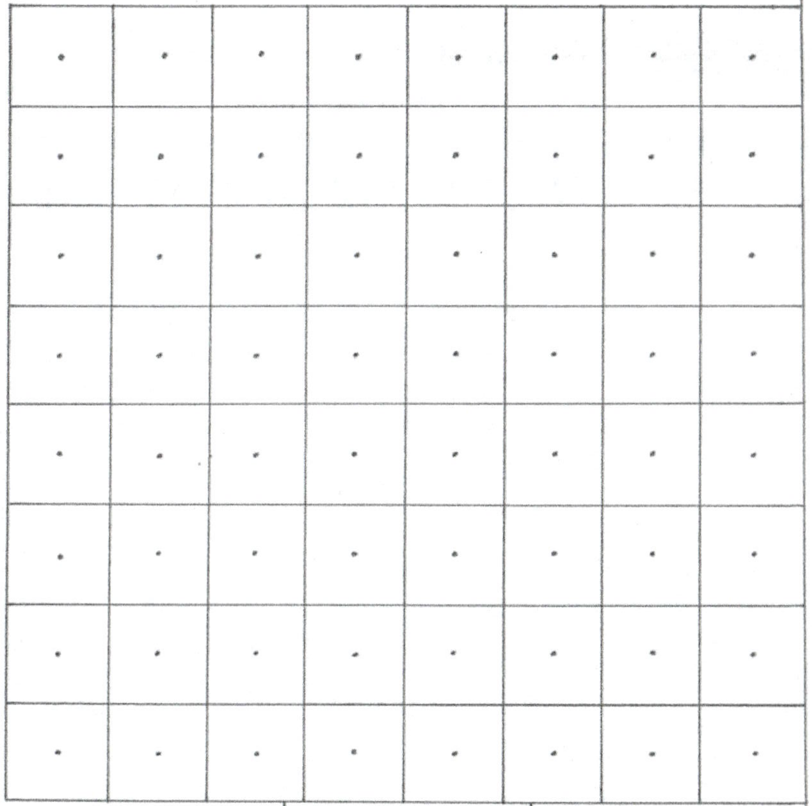

FIG 20

..."When I say that the chair in which I am sitting has remained stationary for the last hour, I mean only stationary in respect to myself & the house. In reality, the Earth's rotation has carried it over one thousand miles & the Earth's course some seventy thousand miles from its previous position..."

—from "777" by Aleister Crowley.

8	58	59	5	4	62	63	1
49	15	14	52	53	11	10	56
41	23	22	44	45	19	18	48
32	34	35	29	28	38	39	25
40	26	27	37	36	30	31	33
17	47	46	20	21	43	42	24
9	55	54	12	13	51	50	16
64	2	3	61	60	6	7	57

FIG 18

In the space of Fig 20 above, connect the BALANCE OF THE PAIRS of the Harmony Square of Mercury i.e Fig 18 shown again on this page. Refer back to Fig 9 which shows the first Balance of the Pairs of the Franklin Harmony Sq.

Introducing
The Ancient
Brahmins
KNIGHT'S TOUR.
Page 31

The Ancient Brahmins' Knight's Tour.

Fig 21 is known specifically as an 8×8 Indian Jaina Majik Square. Instead of drawing a long continuous line from 1→2→3→... 64 as is usually the case, this Knight's Tour will be traced in such a way that every period or grouping of 8 numbers thru-out the tour will also have a sum of 260.

Fig 21

1	59	56	14	2	60	53	15
46	24	27	33	47	21	28	34
32	38	41	19	31	37	44	18
51	9	6	64	50	12	5	63
3	57	54	16	4	58	55	13
48	22	25	35	45	23	26	36
30	40	43	17	29	39	42	20
49	11	8	62	52	10	7	61

To begin the tour follow these specific instructions. In the space of Fig 22, beginning at the No. 2, join a line from 2 → 28 → 63 → 37 → 64 → 38 → 1 → 27 etc.→ and so on until all 64 numbers given are contacted. Note that all these groupings of 8 have a Sum of 260. Join a line from 27→31 and begin again to contact the following: (31, 60, 34, 5, 36, 7, 29, 58) then (44, 15, 21, 50, 23, 52, 42, 13) then (12, 47, 53, 18, 55, 20, 10, 45) then (62, 40, 3, 25, 4, 26, 61, 39,) then (35, 8, 30, 57, 32, 59, 33, 6) then (22, 49, 43, 16, 41, 14, 24, 51) then (54, 17, 11, 48, 9, 46, 56, 19)

And to close The Circuit, join a final connecting line from 19 → 2. ॐ

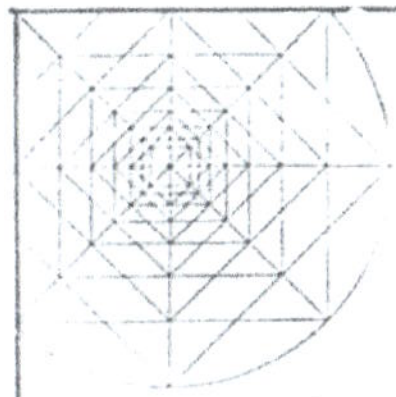

The Ancient Brahmins' Knight's Tour continued

Fig 22

Page 33

Page 34

IF YOU WISH TO FOLLOW
THE PATH OF SAINTS
LEARN TO FORGIVE

1	59	56	14	2	60	53	15
46	24	27	33	47	21	28	34
32	38	41	19	31	37	44	18
51	9	6	64	50	12	5	63
3	57	54	16	4	58	55	13
48	22	25	35	45	23	26	36
30	40	43	17	29	39	42	20
49	11	8	62	52	10	7	61

FIG 21

"FORGIVENESS" is a debatable word. If you Create Your Own Reality & thus There Is No Right or Wrong then there can be nothing to Forgive, but rather just to Let It Be, to ALLOW. This is The Goal of The Knight's Tour. To Be BLAMELESS & NON-JUDGEMENTAL.

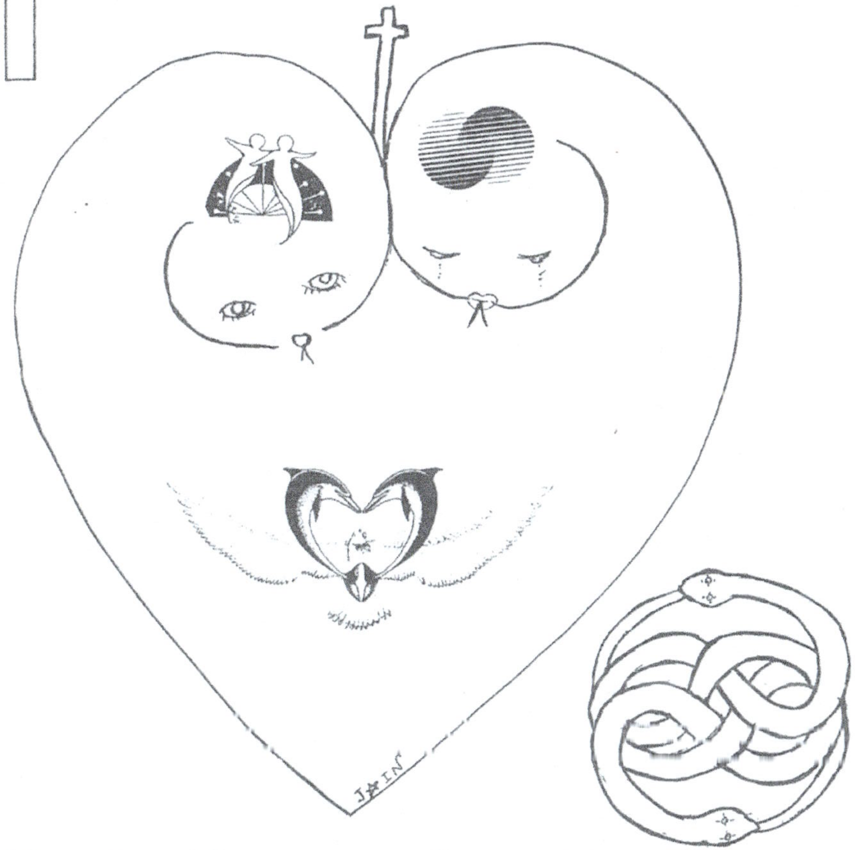

In the space of Fig 23, connect THE BALANCE OF THE PAIRS based on The Knight's Tour: Fig 21. Refer back to Figs 9 & 20 to compare the solutions.

Page 35

Fig 23

.	.	.	.	.	.	.	.
.	.	.	.	.	.	.	.
.	.	.	.	.	.	.	.
.	.	.	.	.	.	.	.
.	.	.	.	.	.	.	.
.	.	.	.	.	.	.	.
.	.	.	.	.	.	.	.
.	.	.	.	.	.	.	.

The Queen's Tour
Page 36

"dancinG into truth" I AM Spirit on Earth &

I dedicate mySelf to channelling love & truth into my Life... I Am a Source of Light in the living body of potential, evergrowing in unity with all, as the universe & I flowww harmoniously in change together... Love is my nature, truth is my will. The essential masculine & feminine principles of God & Goddess in all their forms dwell within me... The ecstasy of their intertwining is the divine spark of life into my deeper nature. I am a wild child, a responsible adult & a wise elder. I encompass within mySelf the wealth of human experience. I welcome the fullness of life with its joy & sorrow as I work & play. Accepting into memory all suppressed experiences, I release them with love & forgiveness. I love & do as I will. The truth of my experience & purity of my heart enable me to govern mySelf. I share a vision with all people building a world based on freedom & co-operation... The Universe flows thru me & I use this power to create, nurture & heal. My love embraces violence & pain & transmutes them into peace & enthusiasm. I respect the autonomy of each & choose not to interfere with pity or harm... I am an artist creating my own reality & providing mySelf with the experiences I need. A divine presence guides my ego & shapes all I think, feel, say & do. I lovingly & willingly surrender to the beauty & perfection of all that I am. I feel my connection with all facets of this universe – the earth, the water, the sky, plants, animals, spirits & other beings.

THE DIVINE WITHIN ME RECOGNISES & APPRECIATES THE DIVINE WITHIN YOU. My mind is a messenger bearing inspiration & intuition. I am sincere, I honor my word, & I fulfill the responsibilities I choose. In perfect love & perfect trust I share the truth within me. Thru the veils of illusion I see clearly the stillness that moves me. Life's experiences reflect my behavior as the web of cause & effect constantly reminds me of my path... WHATEVER I SEND OUT RETURNS MAGNIFIED. I am healthy, happy & free as earth's abundance provides for all my needs. I celebrate the cycles of change & recognise illness, adversity & confusion as opportunities to grow in wisdom & understanding. Realizing the grace of the divine within, I laugh sing & dance in the everpresent now. I am birthing a new magickal being within myself as part of the planetary transformation of consciousness & evolution of homo sapiens into Homo Veritas, true human, Be Come God-dess

I Am, We Are, One, None, All.

Page 37

The Queen's Tour, continued

The Queen's Tour (Fig 24) has the fascinating property that when the squares are numbered consecutively, as they are visited, starting with '1' in the square marked 'S', then it will be a majik square as shown in Fig 25. A 90° rotation is shown as Fig 26. All the Sums of the Columns, Rows & Diagonals are 260.

In the space of Fig 27 neatly construct the QUEEN'S TOUR at 0° + 90° by superimposing the Majik Square Yantras (or Diagrams) of Figs 25 + 26.

61	62	63	64	1	2	3	4
60	11	58	57	8	7	54	5
12	59	10	9	56	55	6	53
13	14	15	16	49	50	51	52
20	19	18	17	48	47	46	45
21	38	23	24	41	42	27	44
37	22	39	40	25	26	43	28
36	35	34	33	32	31	30	29

Fig 25

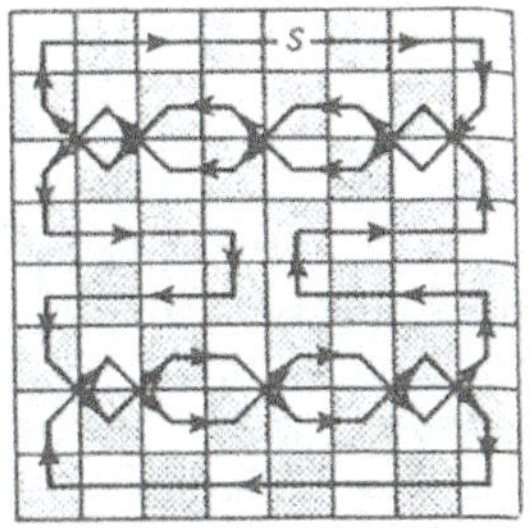

The queen's magic tour

Fig 24

4	5	53	52	45	44	28	29
3	54	6	51	46	27	43	30
2	7	55	50	47	42	26	31
1	8	56	49	48	41	25	32
64	57	9	16	17	24	40	33
63	58	10	15	18	23	39	34
62	11	59	14	19	38	22	35
61	60	12	13	20	21	37	36

Fig 26

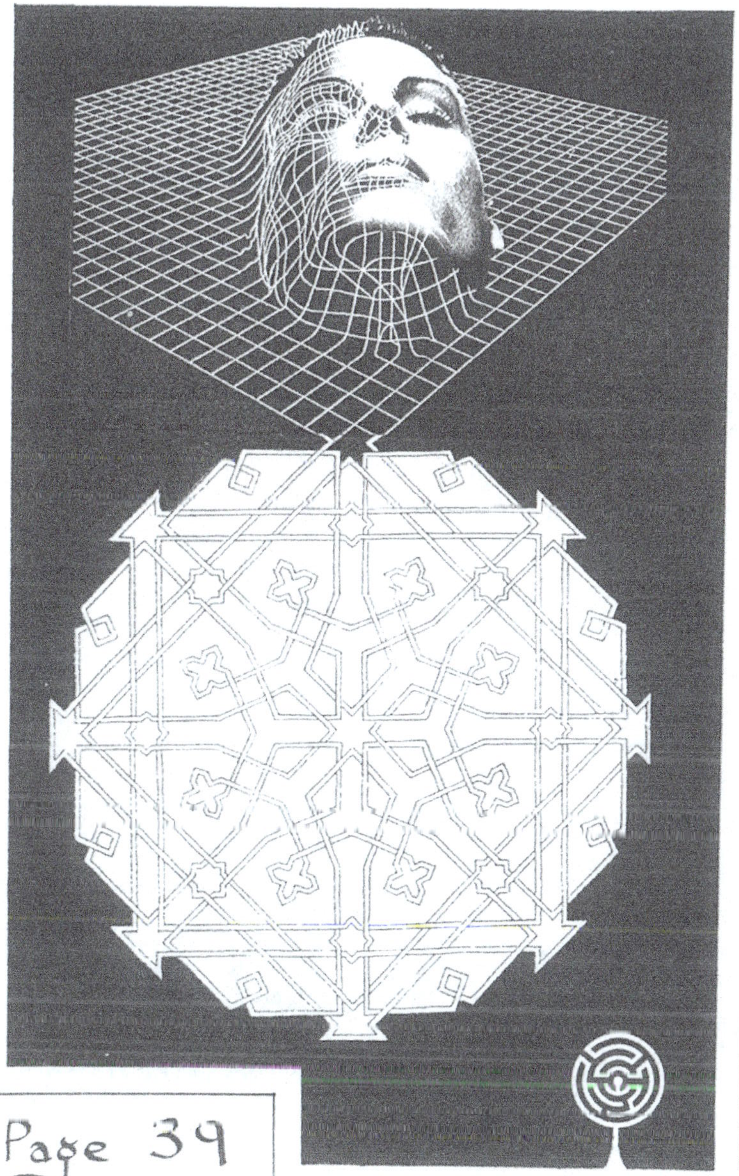

Fig 27

Page 39

Page 40

Deep at the center of my being there is an infinite well of Love.

I now allow this Love to flow to the surface. It fills my heart, my body, my mind, my consciousness, my very being, and radiates out from me in all directions and returns to me multiplied. The more Love I use and give, the more I have to give, the supply is endless. The use of love makes Me feel Good, it is an expression of my inner joy. I Love myself therefore, I take Loving care of my body. I Lovingly feed it nourishing foods and beverages, I Lovingly groom it and dress it, and my body Lovingly responds to me with vibrant health and energy. I Love myself therefore I provide for myself a comfortable home, one that fills all my needs and is a pleasure to be in. I fill the rooms with the vibration of Love so that all who enter, myself included, will feel this Love and be nourished by it. I Love myself therefore, I work at a job that I truly enjoy doing, one that uses my creative talents and abilities, working with and for people that I Love and that Love me, and earning a good income. I Love myself therefore, I behave and think in a Loving way to all people for I know that that which I give out returns to me multiplied. I only attract Loving people in my world for they are a mirror of what I am. I Love myself therefore, I forgive and totally release the past and all past experiences and I am free. I Love myself therefore, I Love totally in the now, experiencing each moment as good and knowing that my future is bright, and joyous, and secure for I am a beloved child of the universe and the universe Lovingly takes care of me now and forever more. And so it is.

61	62	63	64	1	2	3	4
60	11	58	57	8	7	54	5
12	59	10	9	56	55	6	53
13	14	15	16	49	50	51	52
20	19	18	17	48	47	46	45
21	38	23	24	41	42	27	44
37	22	39	40	35	26	43	28
36	35	34	33	32	31	30	29

Fig 25

In the space of Fig 28 connect THE BALANCE OF THE PAIRS based on The Queen's Tour: Fig 25. The Formula is $(1+n^2)$ for $n=8$ which is the first No. 1 + the last No. 8^2 or 64 which sum to 65. This Balancing of the Pairs reiterates what Jesus spoke: "I Am The Alpha & The Omega, The First & The Last."

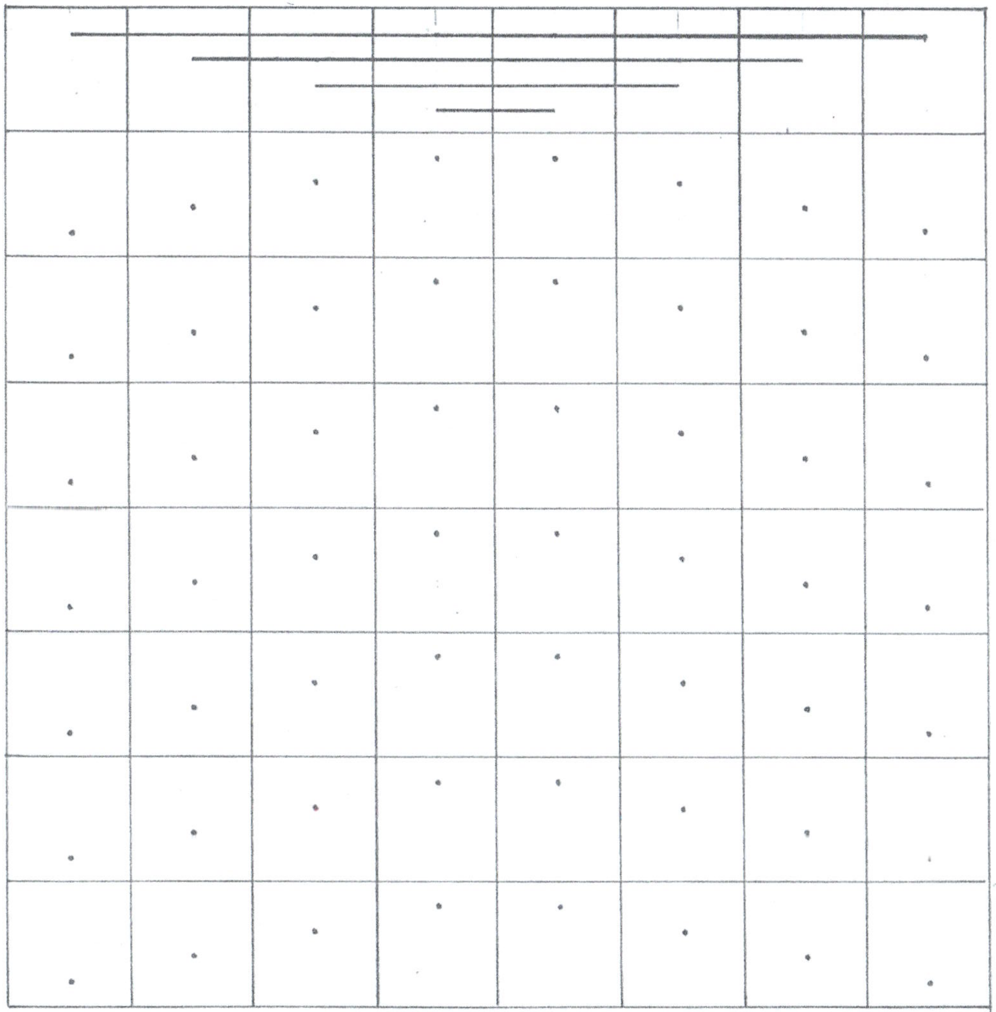

nb: The top row has already been completed.

Fig 28

Page 41

A JAINA HARMONY SQ OF 8×8

The Reverend A.H. Frost, while a missionary for many years in India, of the Church Missionary Society, interested himself in his leisure hours in the study of these squares & cubes, & in the articles which he published on the subject gave them the name of "NASIK", from the town in which he resided. He has also deposited 'Nasik' cubes in the South Kensington Museum (London) & he has a vast mass of unpublished materials of an exhaustive nature most carefully worked out. "Mr Kesson has also treated the same subject in a different way & more popular form in the 'QUEEN' (published in London). He gives them the very appropriate name of "CAISSAN SQUARES", a name given to these squares, he says, by Sir William Jones. The proper name, however, for such squares should rather be "INDIAN", for not only have the Brahmins been known to be great Adepts in the formation of such squares from time immemorial, not only does Mr Frost give his an Indian name but one of these squares is represented over

THE GATE OF GWALIOR.

While the natives of India wear them as amulets, & La Loubère who wrote in 1693, expressly calls them 'INDIAN SQUARES'.

(from: "Magic Squares & Cubes" by W.S. Andrews, 1917).

1	58	3	60	8	63	6	61
16	55	14	53	9	50	11	52
17	42	19	44	24	47	22	45
32	39	30	37	25	34	27	36
57	2	59	4	64	7	62	5
56	15	54	13	49	10	51	12
41	18	43	20	48	23	46	21
40	31	38	29	33	26	35	28

FIG 29

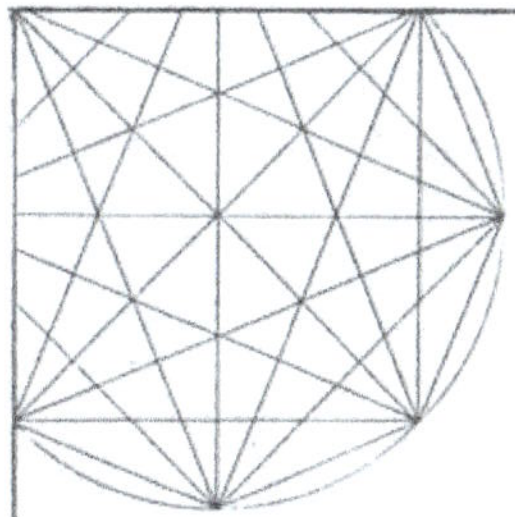

Fig 29 was carved in stone at the entrance to the village of Gwalior, consecrated by the priests since time immemorial, its purpose was to give spiritual protection to the whole of the village.

The Brahmin priests & yogis occupied themselves with creating unique majik sq. combinations as a form of intellectual past-time/challenge.

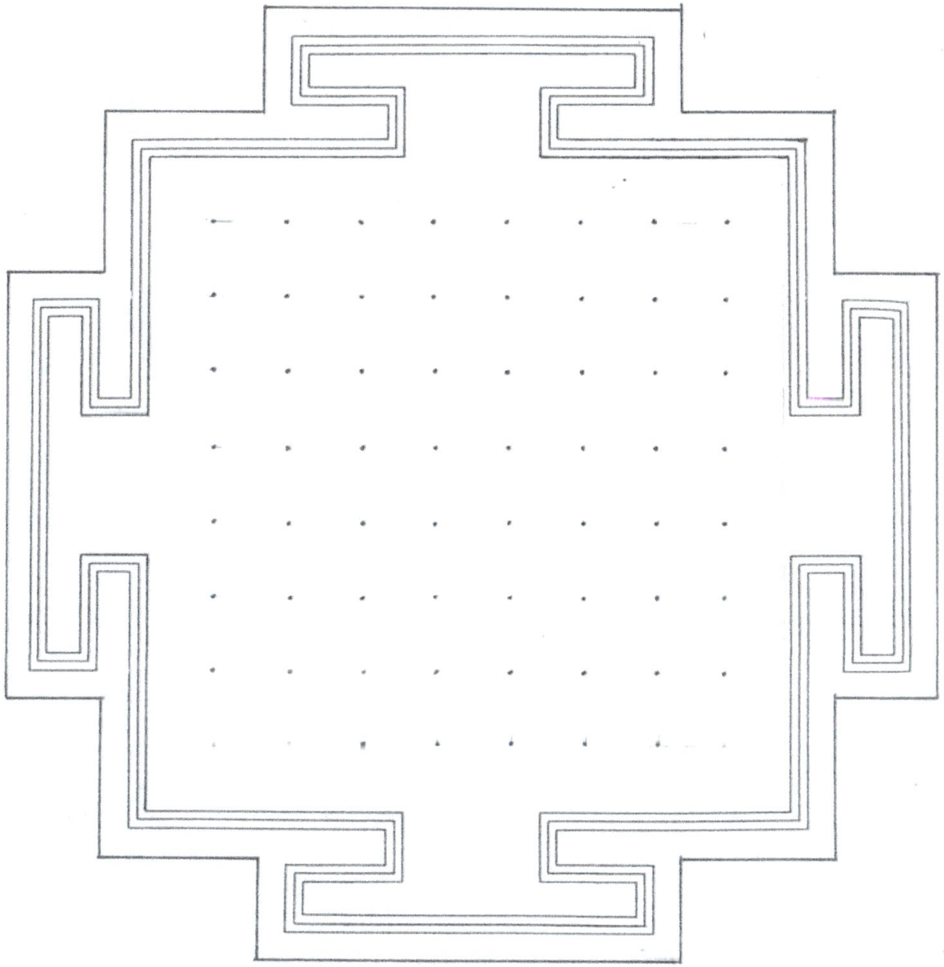

In the Space of Fig 30, construct the Harmony Square Yantra at 0° of the 8x8 Jaina Harmony Sq. shown in Fig 29. That is, just draw a long continuous unbroken line from 1 → 2 → 3 → ··· 64 and close the Circuit by joining 64 → 1. After every Yantra has been drawn check your graphics in the last chapter of "Solutions".

FIG 30

Page 43

Page 44

The PROPERTIES of this Indian Jaina Majik Square of 8×8 are :

- All columns & rows sum 260.
- All bent diagonals, similar to the Franklin Sq of 8×8 sum 260 e.g: 17+55+3+29+48+10+62+36.
- All 2×2 Sub-Squares sum 130.
- The Square remains unchanged if columns are taken from one side & put on the other, or if they are moved from the top to the bottom, or Vice-Versa.

The ancient Jains 2,500 years ago wandered naked, renunciating all material things. Begged for food. They brushed their path so as not to kill a living thing. They died by not eating.

They could not even be potters 'cos that killed organisms. All monies built superb temples so rich with ☩ design. They were the masters of Majik Squares inspired through Meditations.

In the space of Fig 31, construct the Majik Square Yantra of Fig 29 at 0° + 90° + 180° + 270° (i.e: at 4 x 90° rotations). To do the rotations, follow these **steps** :

- a) Rewrite Fig 29 on a separate piece of paper & have it infront of you without moving it until the 4 superimpositions have been completed. When drawn once simply like this it is called "0°" (Nought Degree).
- b) Turn this book anticlockwise until the writing "90°" is infront of you. Then superimpose the Yantra of Fig 30 over itself at 0°. You will notice that some lines have already been drawn, so carry on.
- c) Turn this book again until the writing "180°" is infront of you. Superimpose Fig 30 again over steps a) & b).
- d) Turn this book again until the writing "270°" is infront of you. Superimpose Fig 30 again over steps a), b) & c).

The holy metal → sculpture is typical of the Indian Jain's Art. The Buddha-like figure → is actually an Empty Space & represent's the Eternal Void / SPIRIT.

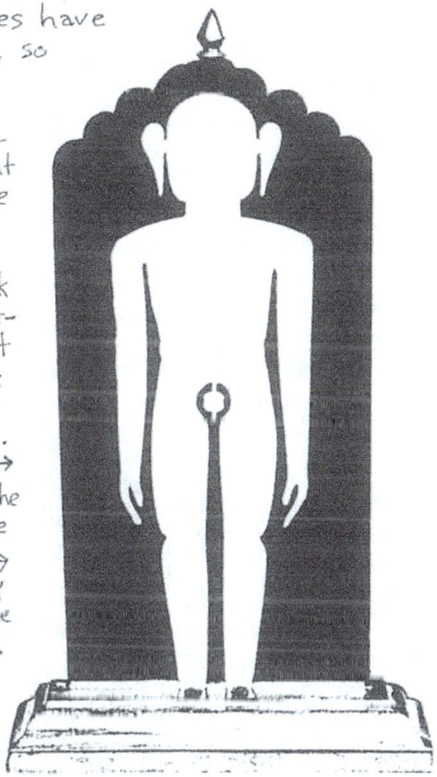

Page 45

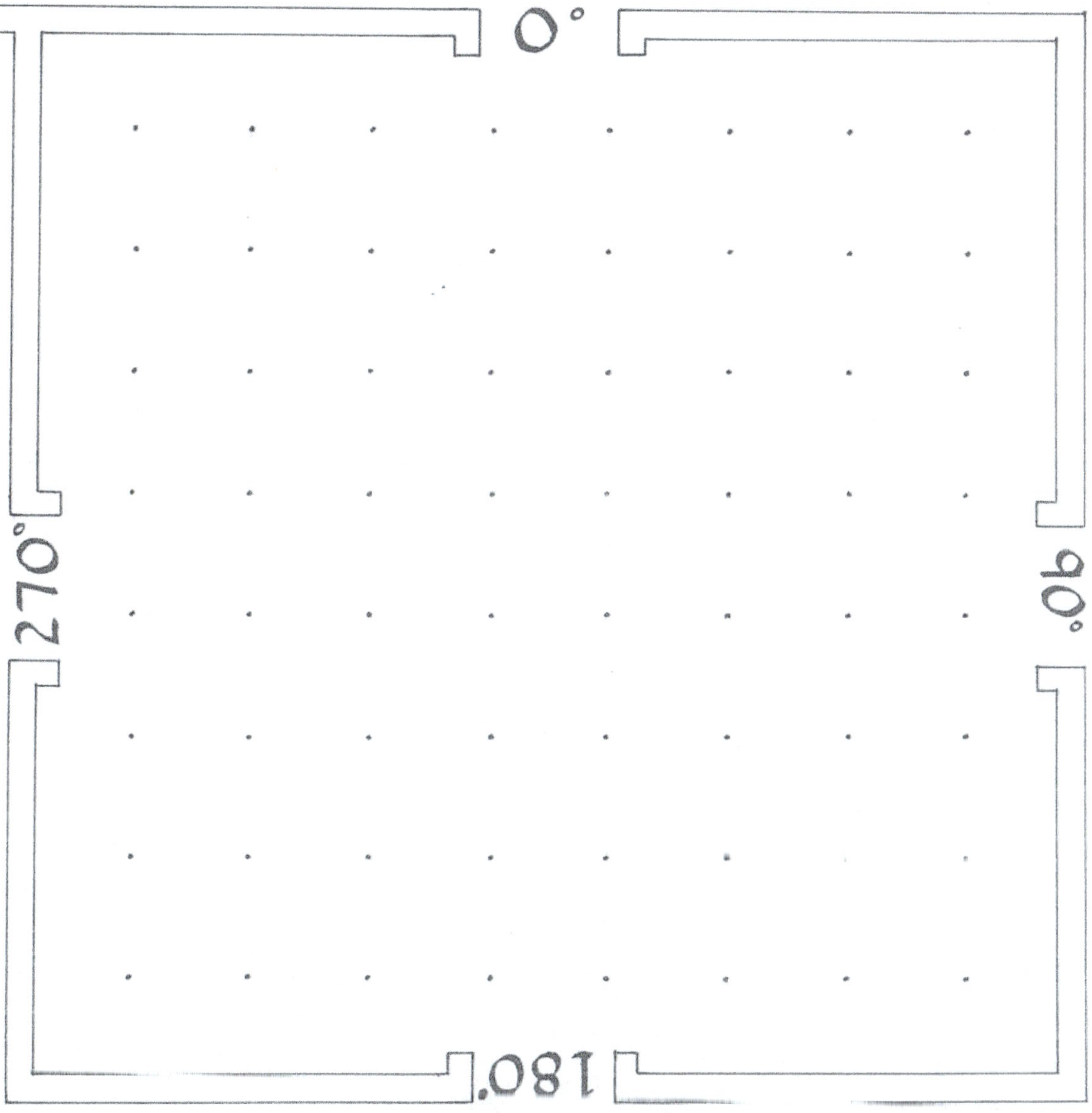

Fig 31

Page 46

THE BALANCE OF THE PAIRS

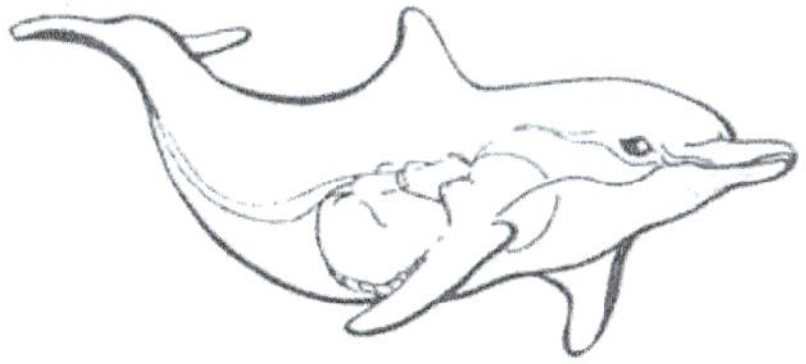

In the space of Fig 32 connect the dots (that are symbolic of the numbers of the 8 × 8 Jaina Majik Square : Fig 29) that when paired have a sum of 65. "65" is the sum of the first + last number of the Majik Square.

The Majik □ Constants e.g C= 260, are secret codes of the unwriteable Names of the Lord in Hebraic Culture.

1	58	3	60	8	63	6	61
16	55	14	53	9	50	11	52
17	42	19	44	24	47	22	45
32	39	30	37	25	34	27	36
57	2	59	4	64	7	62	5
56	15	54	13	49	10	51	12
41	18	43	20	48	23	46	21
40	31	38	29	33	26	35	28

FIG 29

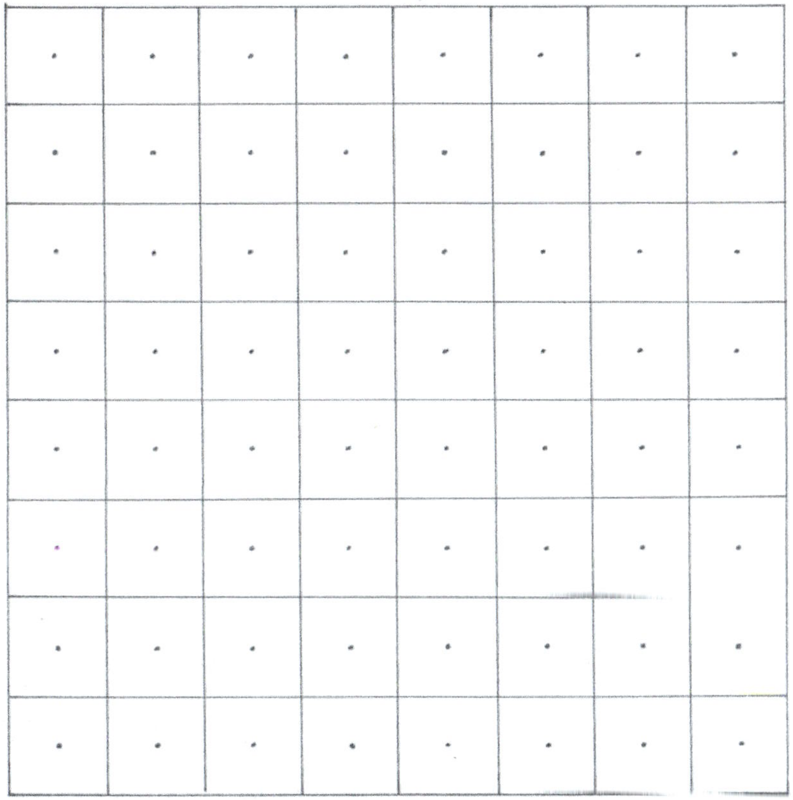

FIG 32

Page 47

Page 48

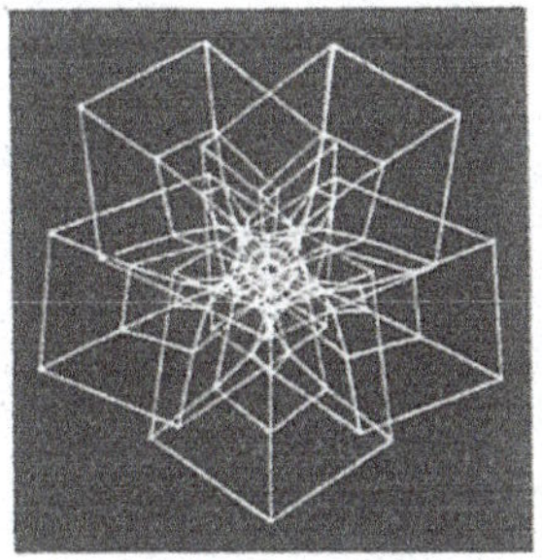

FUNDESIGN

Create these fascinating designs which helps develop concentration

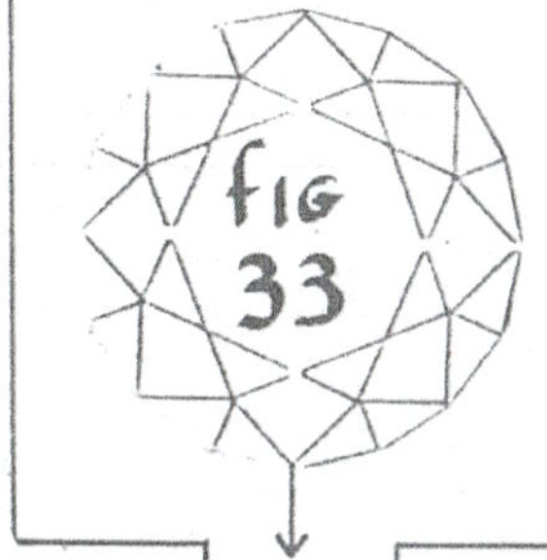

1	32	40	57	56	41	17	16
2	31	39	58	55	42	18	15
3	30	38	59	54	43	19	14
4	29	37	60	53	44	20	13
61	36	28	5	12	21	45	52
62	35	27	6	11	22	46	51
63	34	26	7	10	23	47	50
64	33	25	8	9	24	48	49

This majik Square of 8×8 was composed by L.S. FRIERSON.

Here, like the Ben. Franklin Square, it has bent diagonals summing 260. eg 1+31+38+60+53+43+18+16 = 260 etc.

FIG 34

In the space of Fig 34 construct the Harmony Square of 8×8 Yantra composed by L. S. Frierson as shown in Fig 33 on the previous page.

Page 49

Shown below, FIG 35 is a typical

TIBETAN EYE EXERCISE. The numbers are to show the course your eyes take as they travel from the inner circle to each outer circle. Focus on the circle in the centre. Now travel with your eyes: once around the centre circle & outward along the dots until you reach an outer circle (once around it) & follow the dots back to the centre (once around again) & out to another circle until you do them all.

"When you do this with reverence or in a quiet meditative space it will naturally improve poor eye conditions like myopia.

Fig 35 is designed to inspire you to create your own 'Eye-Exercise Mandala' whether it be hand-drawn free-formly or constructed straight-linedly like the design on your left. (a M.Sq of 7×7 at 8 × 22.5°).

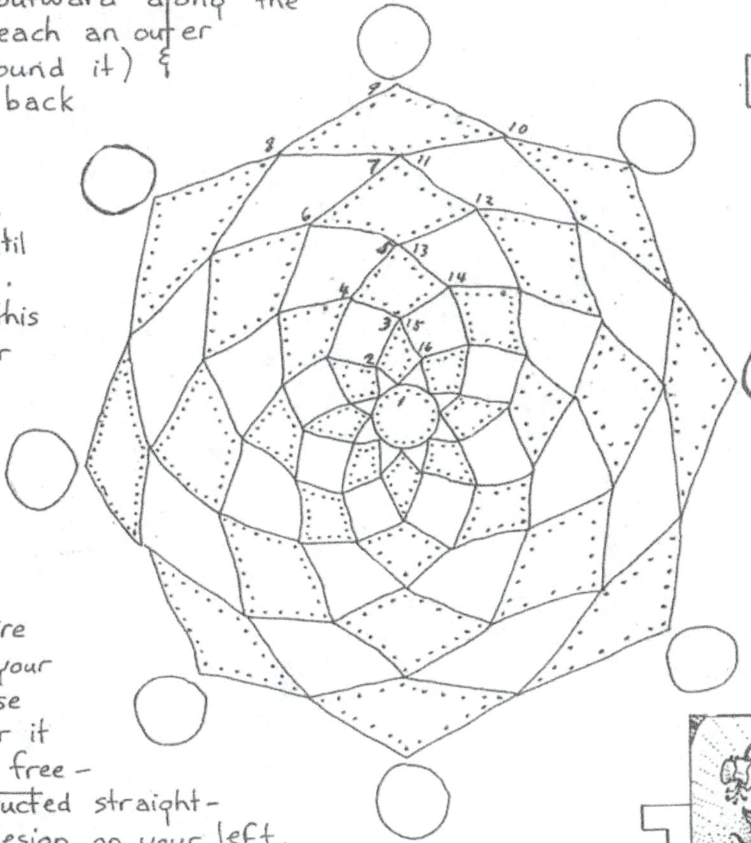

FIG 35

VASTU PURUSHA MANDALA

COSMOS DWELLINGS

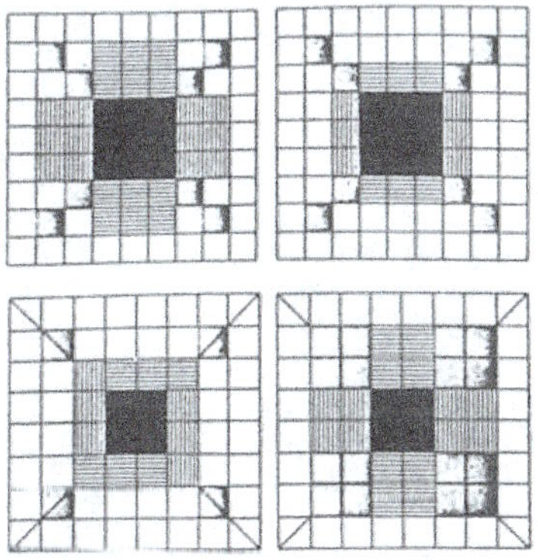

Four types of Vāstu-Purusha Mandala. The centre is the seat of the principal deity, with auxiliary deities extending to the outer borders

FIG 36

The Vastu-Purusha Mandala (Fig 36) is basically
a square of squares bu[illegible] be converted into other
primal shapes. The simpl[illegible] ement consists of 64
(8×8) squares in which [illegible] central zone of 4 divis-
ions is dedicated to the f[illegible] y, Brahmā. Other more
highly complex types can [illegible] ed by gnomic progression into
1024 squares. When [illegible] used architecturally its central zone
locates the temple [illegible] This ritual diagram of Fig 36 is an
"ideogram", while [illegible] materialization of the concept it embod-
ies. Around the nucle[illegible] are designated as seats of divinities,
with specific reference [illegible] ections of space. These are surrounded
by another 32 divinities [illegible] al bodies (28 lunar & 4 that preside
over solsticial & equino[illegible]. Thus this simple graph-like diagram
not only represents th[illegible] f the directions of the compass but
has astronomical con[illegible] iding a chart of cyclical revolution
of day, month, year [illegible] andalas, of which Hindu manuals
of architecture pr[illegible] ns, are the earliest type of
architectural yantras, contribu[illegible] bstantially to the rhythms, designs
[illegible] conceptual basis of the Hindu [illegible] e.
ct a certain cosmic order, while [illegible] se
a town's orientation. Before a temple is
[illegible] rified & consecrated, & the Vastu-Purusha
Thus the building draws its power from the

These concentric [illegible]
in planning a town [illegible]
built, the ground is [illegible]
Mandala is ritually drawn c[illegible]
Vastu-Purusha (Supreme Principle [illegible] o lives
at its base & converts, thru B[illegible] presence,
the plan of existence (Vastu [illegible] into the
shape of the Purusha, in whos[illegible] the
temple is set u[illegible] ays, the
symbol of Ganesha which is the
Swaztika (or majik Square of the
Sun) is laid under the first
Corner Stone, for blessing.

Fig 37

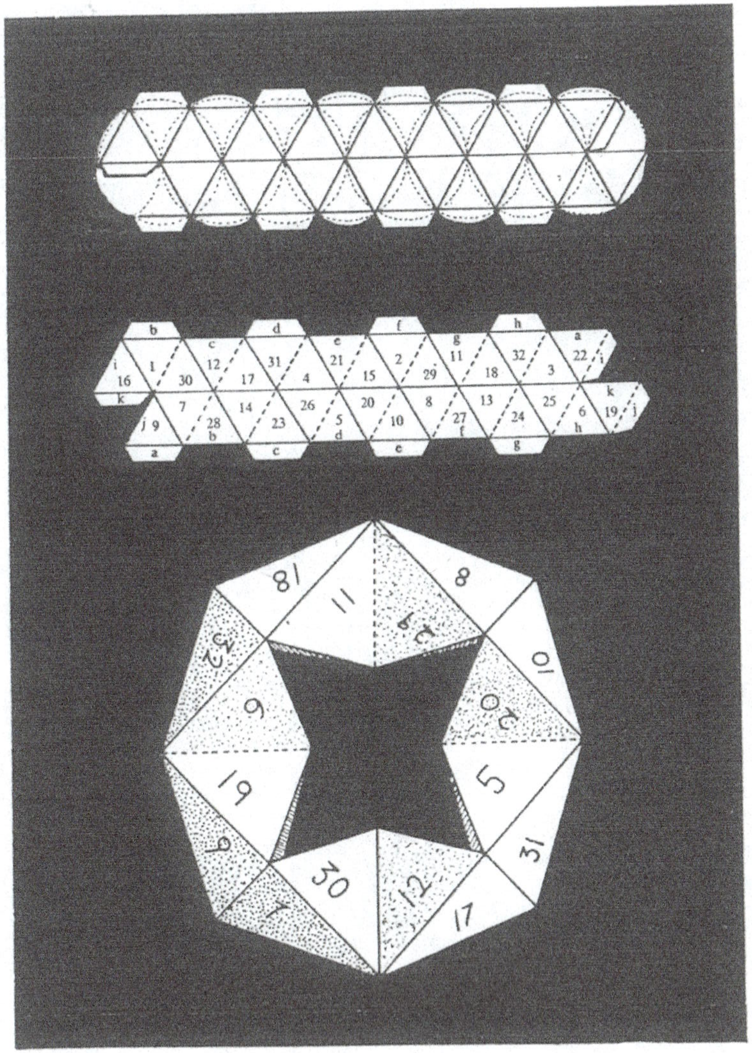

A Rotating Magic Ring of 8ight Tetrahedra.

- On a large piece of cardboard construct the net shown in the top diagram of Fig 37, so that it is 63 cm, & setting your compass at a radial length of 7 cm to make 10 circles.
- The next diagram shows how the tabs have been carefully labelled & the equilateral triangles have been numbered. The edges of the triangles are 7 cm.
- The 3rd diagram shows a bird's eye view.
- Cut out, then fold the dotted lines away from you & the black lines in the opposite direction. Crease thoroughly. Glue the tabs to their corresponding edges i.e: tab 'b' joins to edge 'b', etc. The 8 tetrahedra will come together in the centre & when flexed or rotated, forms an attractive model. WHAT IS THE SUM OF THE NUMBERS ON EACH TETRAHEDRON?

Unification is carried out in action, in devotion, and in contemplation. Here a prayer is written out in the figure of a man who wears a crown. Thus, while the Kabbalist prays, he becomes aware of the head, body and limbs both of himself and of Adam Kadmon, the primordial man . He shifts his consciousness from the personal to the Divine, thereby bringing himself into union with Adam Kadmon, of whom he is a reflection, and who is in turn the reflection of the Divine. Thus a human individual aids God to behold God. (Divine image in prayer from ms. Hebrew prayer book, France or Germany, 13th–14th c. British Library, London.)

(This Magic Rotating Ring has a magic No. (Constant) of 132).

Consider this doodle. The purpose of this doodle on this page in this book is to express to you to be Loose. That's the total message of The books of Majik Squares; its not just to copy & do geometrically what you are instructed, but to be imagineering, different, Xplorational, taking risks, developing new modes of thinking, meditating, becoming, endlessly becoming, educrealional, becoming artistically mathemagical, ingenuously dyslexical, becoming Loose. In the words of a musical group in the 60's known as "Cheech & Chong" : "If you are not loose, get Loose; If you are already loose, get Looser." Just observe how loose the lines are of this contortioned doodle, how the body is bezerkly out of proportion, how mindless it is, yet it is dancing divinely spastic & All-knowing.
Jain
JAIN
ON THE ART OF BEING HERE NOW.
I AM
JOURNEY OF AWAKENING:
Page 53

Page 54

Fig 38

"A Spiral of Triangles Manifesting thru an 8×8 Matrix of Squares."

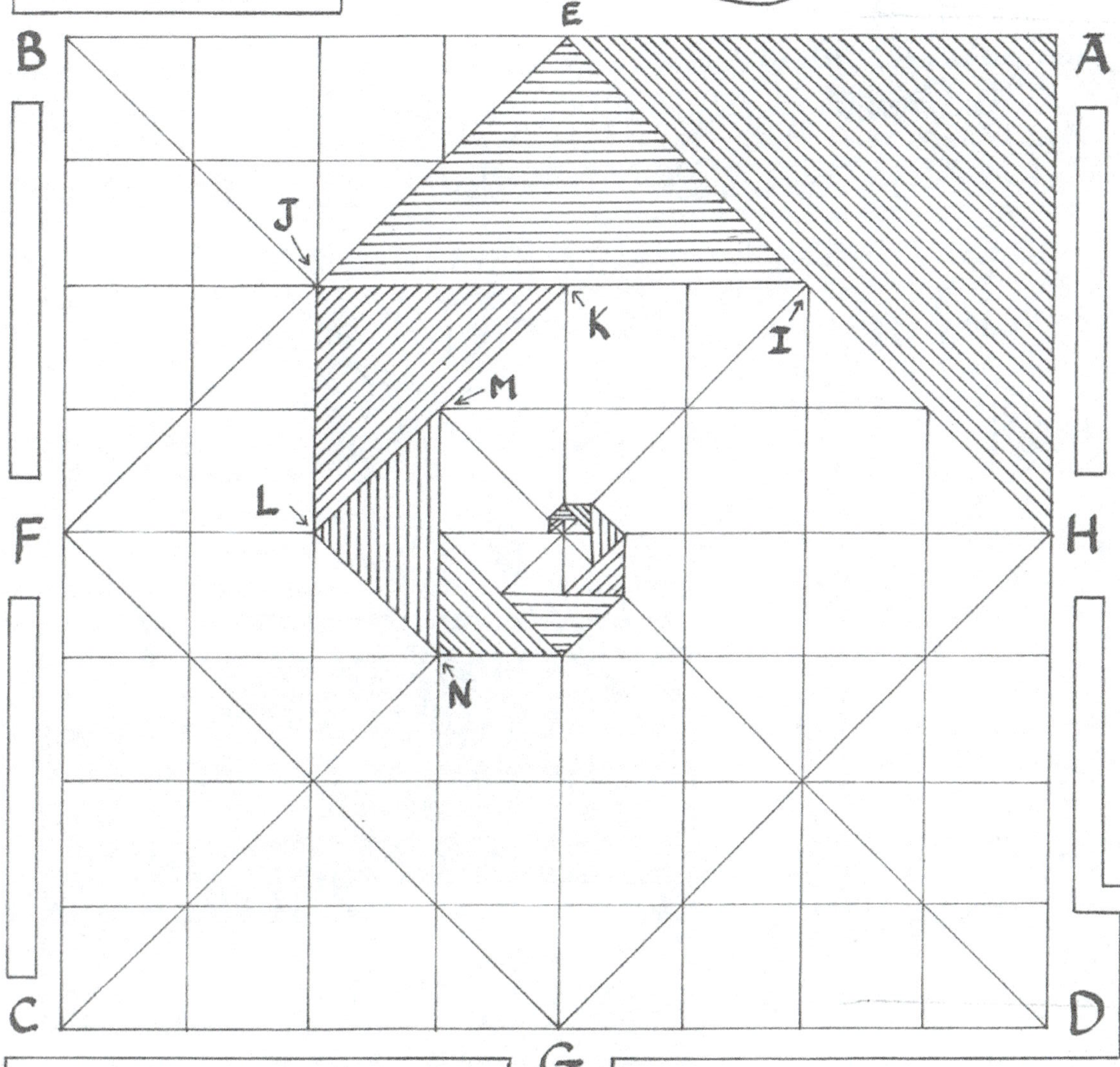

- Method of Construction : Make a large sq. ABCD. Draw in diagonals BD & AC. Let EFGH be a diamond in the square. Color-in the Right-Angled Triangle AEH. Let "I" be the midpoint of the hypotenuse EH (the longest length) & this determines the diminished size of the next triangle IEJ. Let "K" be the midpoint of JI & then form the next diminished Δ KJL. Keep following this procedure until the Δs are so small that inaccuracy stops the process.
- Question : Focus upon the centre of the whole square ABCD. Does the Spiral of Diminishing Triangles ever contact the centre or does it forever spiral about it as it continually approaches the centre ?
- Assignment :

On the following page, in the space of Fig 39 a 9×9 grid of dots has been given to you to form an 8×8 matrix of squares. this is to be drawn in lightly in pencil. Follow the above Method of Construction as outlined graphically for you in Fig 38 until the centre is approached.

- Suggestion : Color in your spiral of triangles in a rainbowic order, i.e : Color in Δ AEH Red, then IEJ Orange, KJL Yellow, MLN Green, followed by blue, indigo, violet.

Page 55

"SPIRAL OF TRIANGLES" Fig 39

MOSAIC LAW

FIG 40

Fig 40 depicts the Blue Mosque of ISFAHAN, 8th Century, fretted and inlaid window ('fret' means an interlaced angular design of bands within a border). The intricate decoration carries on from the Byzantine tradition. This was given added impetus by the strict observance of Mosaic Law forbidding iconographic reference to men/women & animals. Instead, floral & geometric patterns were developed to the ultimate degree. Thus the ancient Moslems were master Iconologists (the branch of knowledge concerned with pictoral or sculptural representation). More precisely, they were Majik Square Iconologists in that specific M. Sqs which they created did transform number sequences into an exquisite ArtForm. Having made two visits to Arabia, I can clearly state that the ancient mosque Temple Art is based on M. Sqs but in a very elaborate & secretive way. The manner in which I have developed M. Sq. Art by simply drawing a long continuous line of electricity from beginning to end, & rotating & superimposing the same, is kindergarten to them. Tho' their methods have been explained to me. I still have not fathomed their artistic rituals. It is an ancient process taught from master to student individually & orally & cannot be expressed easily in text.

The Koran stated boldly that People were forbidden to draw human or animal images because the Lord/Lady created man/woman in the likeness of his/her Divine Self. Thus they became the Masters & Mistresses of Mathematics, Abstracts, & the Majik Square Geometries.

Moses.

As a Custodian of Majik Squares, Phi Spirals & the Mystical Squaring of the Circle) I know that these Yantras will help heal the Earth. These eternal yantras which symbolize "ORDER AMIDST CHAOS" or "Unity Beyond Division" serve to act as Transistor Circuits or Channels conducting Psychic Essences. For a practical use, place your medicine bottles upon the centre of a M.Sq. yantra. M.Sqs. Amplify & Clarify the Vibration of anything you choose to use, just like a Pyramid does. (n.b. 'PYra' = Fire & 'mid' = centre).

Page 58
Chapter NINE

WORLD INFORMATION SERVICE
Now the secret is out.
INTRODUCING
The Magic Power of
THE
ESSENE
MAGIC
SQUARE
OF
ENTER ANOTHER WORLD
Harmony
DESIGNS FOR THE NEW DECADE
JAIN
RARE WORKS
Page 59

37	78	29	70	21	62	13	54	5
6	38	79	30	71	22	63	14	46
47	7	39	80	31	72	23	55	15
16	48	8	40	81	32	64	24	56
57	17	49	9	41	73	33	65	25
26	58	18	50	1	42	74	34	66
67	27	59	10	51	2	43	75	35
36	68	19	60	11	52	3	44	76
77	28	69	20	61	12	53	4	45

FIG 41

Instructions–

Select your favourite color pencil & very lightly shade in all the odd numbers of the Essene Magic Square of 9x9 as shown in Fig 41. This will demonstrate an underlying symmetry between the odd & even numbers. Constantly meditate upon the divine Plan of your Life.

MAGIC SQUARE MISSION

In the space of Fig 42 carefully construct the 9x9 Essene Magic Square Yantra at 0°: Like usual, always connect the last number to the first so it becomes a closed electrical circuit ∞ flowing.

THE ESSENE MAJIK SQUARE OF 9 X9

Page 61

FIG 42

Fig 41 is The Essene Majik Sq. of 9x9. According to traditional Qabalahistic teaching it represents the archetype of the Moon. "I am the light of the moon." (Bhagavad-gītā 7.8) The Sum of all the Columns Rows & Diagonals = 369. The Sum of all the Numbers = 3,321. The Sum of all central crosses also = 369 eg: 31 + 81 +41 + 1+51 + 49 + 9 +73 +33

MOON

This Mystical Sum of the Cross was why early Christian scrolls showed Magic Sqs. in their index + magic Sq. Parables thru-out "The Gospel of The Holy Twelve" in the author's Library.

Fig 41

37	78	29	70	21	62	13	54	5
6	38	79	30	71	22	63	14	46
47	7	39	80	31	72	23	55	15
16	48	8	40	81	32	64	24	56
57	17	49	9	41	73	33	65	25
26	58	18	50	1	42	74	34	66
67	27	59	10	51	2	43	75	35
36	68	19	60	11	52	3	44	76
77	28	69	20	61	12	53	4	45

Fig 41 is repeated here again so that on the following page you can begin to construct the 9x9 Essene Majik Square Yantra at 0° + 90°. (If you were only presented with Fig 41 & not both Figs 41 & 43 there would be more labour involved for you to construct the 90° rotation by having to rewrite Fig 41 on a separate card & then rotating this book at 90°).

5	46	15	56	25	66	35	76	45
54	14	55	24	65	34	75	44	4
13	63	23	64	33	74	43	3	53
62	22	72	32	73	42	2	52	12
21	71	31	81	41	1	51	11	61
70	30	80	40	9	50	10	60	20
29	79	39	8	49	18	59	19	69
78	38	7	48	17	58	27	68	28
37	6	47	16	57	26	67	36	77

Fig 43

→ ↓ ← ↑

Fig 43 is the 90° Rotation of Fig 41. In the space of Fig 44 superimpose Fig 43 over Fig 41 to create the Essene Majik Square of 9 x 9 at 0° + 90°.

Page 63

Ancient and Free

Fig 44

It is when you start to rotate the Majik Squares that the apparent CRYSTALLINE nature of these number geometries begin to manifest. These ancient yantras are powerful tools to meditate with & it is beneficial to redraw them, enlarge them & paint them as living extensions of your aura. Fig 45 shows the same Essene 9x9 Majik Square Yantra drawn at 4 x 45° i.e: at 0° + 45° + 90° + 135°. The flash of lightning symbolizes how your Resolutions & Ideas, when given Life, confidence & devotion, can manifest.

As you can see on the following 2 pages Fig 46 is the same yantra of Fig 41 drawn at 8x 22.5°. Meditate upon the centre & notice how, from the pure straight-lined geometry, clear visible RINGS of Energy or ripples of concentric circles emanate from the heART Centre.

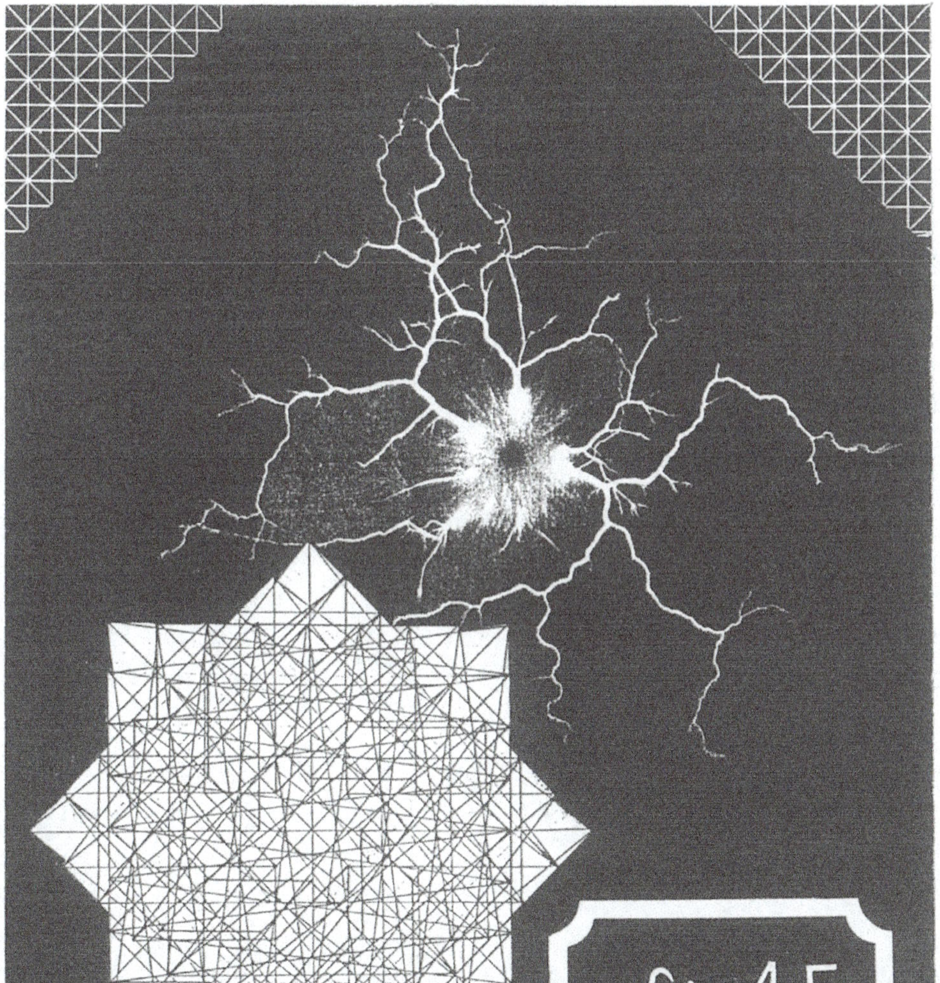

fig 45

I NOW MANIFEST A MODERN MANDALA PUBLISHING HOUSE AMIDST RAIN-FOREST GARDENS, AN INSTITUTE KNOWN AS : "F.R.E.E.D.O.M.S". WHICH IS AN ACRONYM FOR : "FOR RESEARCH EXPRESSING ESSENTIAL DATA OF MAJIK SQUARES."

'AN INTERVIEW WITH JAIN BY JAIN". 1984.

Question : What is your Life's Purpose?

Answer : My Life's Purpose, the heART of F.R.E.E.D.O.M.S., is to CODIFY & establish a functioning system of Majik Square Knowledge & Science in Cyber-Space, stone & text. It is to sieve thru the mass of pedantic data that is scattered around the world & that has come my way, to re-present it in its simplest & artistic form.

I now attract to me all books from all cultures from all time, that have mention of "Magic Squares" so as to house this data in the "F.R.E.E.D.O.M.S." Institute Library: Mystery School.

A Computer Program has been developed in North NSW for the generation of accurate & perfectly constructed Majik Square Yantras. this means that by 1996 all diagrams hand-drawn like Fig 46 will be perfected via the electronic revolution.

THE WORLD'S GREAT LIBRARIES

Page 66

This page shows Fig 46 again but in an enlarged form. Being hand-drawn 9 years ago then fotocopied, it is somewhat inaccurate & of poor quality. I deliberately present this image as a call to Computer-minded Brothers & Sisters who can professionally tune-into these designs with 3-D holographic insights. Fig 46 will suffice for now for the purpose of this book is to stimulate the ingenium or latent creativity of the child. This burst of inspiration is likened to "Savitri" or The Internal Sun, which is an inner feminine principle that is beginning to re-ARISE, world-wide. This material links up the left & right hemispheres of the Brain, the West with the East, the logical with the Intuitive the beginningless Beginning with the endless End. ॐ

FREEDOMS:

ESSENTIAL DATA (For Graphics, **OF MAGIC SQUARES**

FUTURE TECHNOLOGIES.
PROPRIETARY LIMITED

FURTHER INFORMATION

Fig 46

In the space of fig 47 join all the pairs of Nos. in fig 41 that sum $1+9^2$ i.e: 82. To do this, rewrite fig 41 on a separate card & have it infront of you.

When a M.Sq has its pairs Equidistant from the centre it is known as an ASSOCIATED M. SQUARE.

Page 67

THE BALANCE OF THE PAIRS

fig 47

"In this galaxy, the highest totality of Life possible is to simultaneously occupy three body vehicles in as many as NINE time cells... the Nine dimensions in our local universe ...permitting the Christed OverSelf to function..."

"Top View of the Spinnings of your MER·KA·BAH" (Refer to page 266).

HOW TO CONSTRUCT The ESSENE HARMONY SQUARE OF NINE By The METHOD OF TRANSPOSITION.

Page 69

The CONSTRUCTION of M.Sqs. has fascinated Brahmins of India, Sages of China, Monks of Lebanon, & Lamas of Thibet for thousands of years. Fig 48 is part of a step-by-step account of this inherited knowledge.

The METHOD OF TRANSPOSITION is used for the construction of the odd-numbered majik squares i.e for n = 3, 5, 7, 9, 11 etc. (By definition, any M.Sq constructed in this manner has been termed, by the author, as an Essene Harmony Square due to biblical references that will be revealed in Chapter 11).

There is a basic formula for The Initial Boxed Framework that contains all the numbers in arithmetical order (Fig 48). The Formula for the Initial Boxed Framework is $(2n-1)^2$ or $(2n-1)\times(2n-1)$ ∴ For a M.Sq. of 9, the frame size is : $(2\times9-1)\times(2\times9-1)$ for n = 9. Thus we construct a 17 × 17 framework as shown in Fig 48. Beginning with "1" in the centre of the bottom row, commence to write all the numbers from 1 to 9^2 in a diamond-shape. The first 18 numbers have been entered. Complete it.

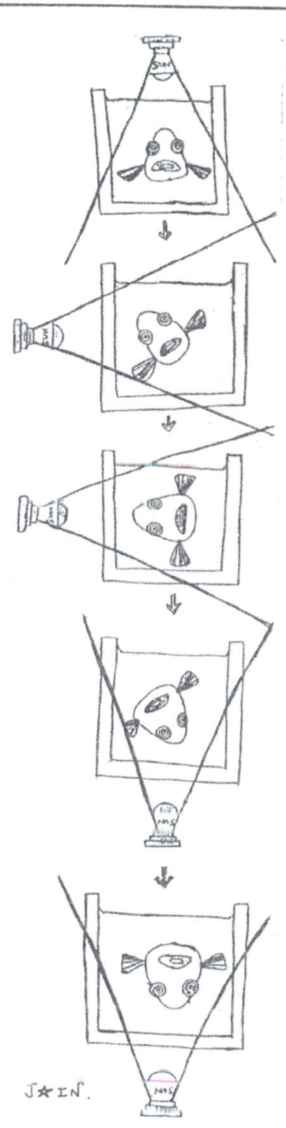

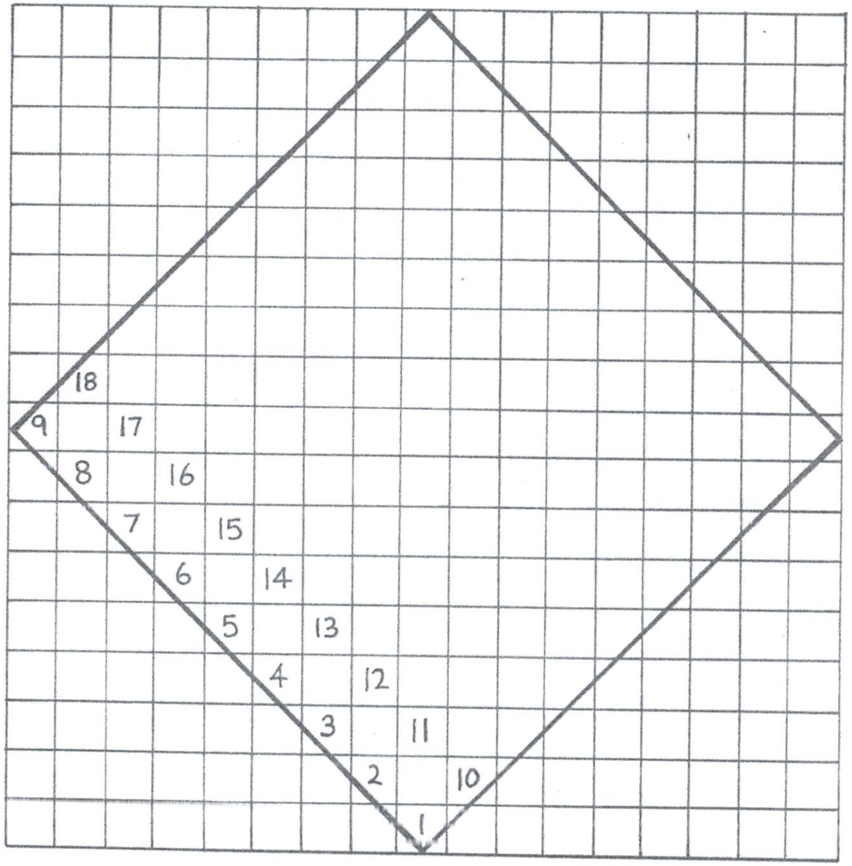

FIG 48

← This doodle is called "Five Steps to Cosmic Consciousness". The word "Transposition" literally means "to move across". Here the fish has kept itSelf in alignment with the Sun source & ultimately has 'moved across' to Upsidedown Reality.

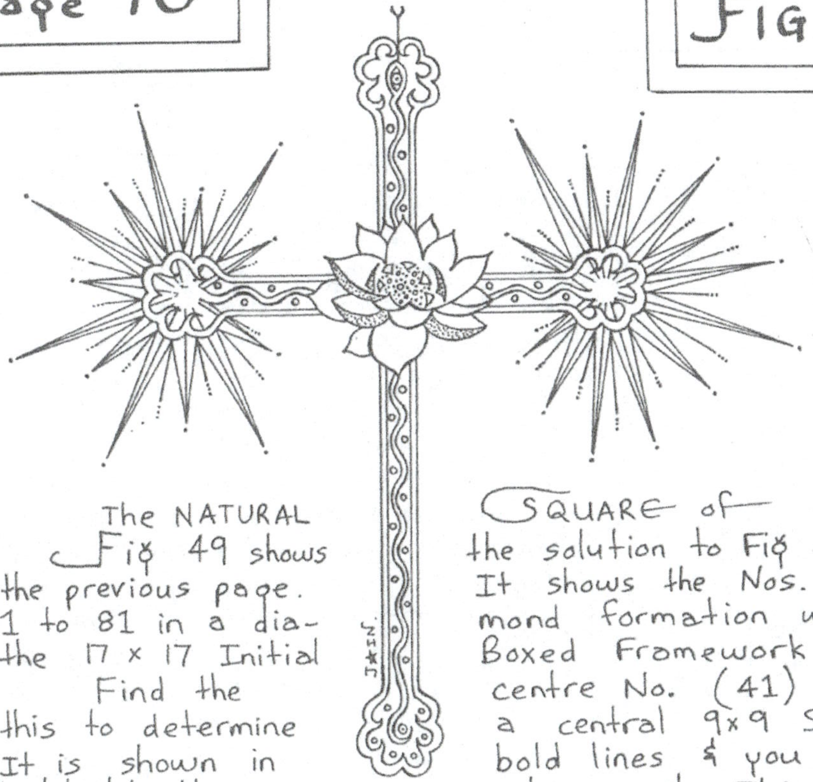

The NATURAL SQUARE of Fig 49 shows the solution to Fig 48 on the previous page. It shows the Nos. from 1 to 81 in a diamond formation within the 17 x 17 Initial Boxed Framework. Find the centre No. (41) & use this to determine a central 9x9 Square. It is shown in bold lines & you can highlight this in color pencil. This inner 9x9 square will contain the imminent Essene Majik Square of 9x9 when all the vacant cells have been filled or occupied by the 4 transposed triangular groups of numbers that lie outside of the central 9x9 Square. To transpose these numbers that lie outside, simply count 9 cells towards the Centre. Each number will eventually fill a vacant cell within the central 9x9 Square. The resulting Majik Square is shown on the next page in Fig 50. To construct an 11x11 M.Sq, you would be required to transpose all outer Nos. 11 cells to the Centre.

FIG 49→

								81								
							72		80							
						63		71		79						
					54		62		70		78					
				45		53		61		69		77				
			36		44		52		60		68		76			
		27		35		43		51		59		67		75		
	18		26		34		42		50		58		66		74	
9		17		25		33		41		49		57		65		73
	8		16		24		32		40		48		56		64	
		7		15		23		31		39		47		55		
			6		14		22		30		38		46			
				5		13		21		29		37				
					4		12		20		28					
						3		11		19						
							2		10							
								1								

Page 71

TOWARD THE ONE

FIG 50

Here is the completed Essene Majik Square of 9x9 formed by the ancient Method of TRANSPOSITION. This enables you to construct any odd-numbered Majik Square. To test your understanding you will be required to construct the Essene Majik Square of 11 x 11 in Chap 11. This method is one of many other methods known. May this technique inspire you to develop your own ideas & see the patterns within the patterns.

								81								
							72		80							
						63		71		79						
					54		62		70		78					
				45	4	53	12	61	20	69	28	77				
			36	76	44	3	52	11	60	19	68	36	76			
		27		35	75	43	2	51	10	59	27	67		75		
	18		26	66	34	74	42	1	50	18	58	26	66		74	
9		17		25	65	33	73	41	9	49	17	57		65		73
	8		16	56	24	64	32	81	40	8	48	16	56		64	
		7		15	55	23	72	31	80	39	7	47		55		
			6	46	14	63	22	71	30	79	38	6	46			
				5	54	13	62	21	70	29	78	37				
					4		12		20		28					
						3		11		19						
							2		10							
								1								

THE UNI-CENTRED SUFI SQUARES of 5, 7 & 9

A decad ago I found a quaint 5x5 m. sq. in The Brittanica. It had at its centre, the Number One, whereas no other m. sq. in my collection had this unique property. By studying its construction I was able to devise similar m. sqs. of 7x7, 9x9 etc. Fig 51, this **UNIT-CENTRED 5x5** M. Sq. was labelled "Islamic", so I decided to call these types of M. Sqs "SUFI". (The Whirling dervish Sufis embrace & honor all people from all religions & openly chant their sacred universal songs). This Oneness is Mystically symbolized by the Centre Cell being occupied by the Number One (or as an empty cell).

23	6	19	2	15
4	12	25	8	16
10	18		14	22
11	24	7	20	3
17	5	13	21	9

FIG 51

*ALLAH is the One God

The Uni-Centred 9×9 Sufi-Square of 9×9 whose Constant = 369 is not going to be given to you, but rather you will construct the 7×7. By studying Fig 52 you will understand the constructional pattern of the mystic Islamic Uni-Centred 5×5 Majik Square. Here is a step-by-step account of how to arrive at Fig 51:

The completed M.Sq. of 5×5 (Fig 51) will be the central square marked ABCD in Fig 52. The 8 surrounding squares are called Imaginary or Auxillary squares & any number that lands in one of the squares must be mirrored back into the central main square ABCD. To begin, observe that the No. "1" is written in the Centre. Now it is your choice where "2" goes. Shown here is "THE KNIGHT's move" 2 cells up & 1 cell to the Right. All moves must be in this same direction until the last No. 25 is entered (except for the multiples of 5 which will be explained). The next No. "3" moves out of the main square & into an imaginary square FGAB. Imagine that you can pick this square up & superimpose it over ABCD. This will give you the No. "3"'s position in ABCD. Keep referring to Fig 51. Then the No "4" ends up in the square BIJD. No "5" ends up in the square FGAB. At every multiple of 5 i.e: 5, 10, 15, 20, there is what you call a BREAKMOVE. Again it is your choice whether you move 2 cells or 3 cells down. Here, we have chosen 1 cell down & this must be used thru-out. Thus the No. "6" ends up in the square CDML etc. Carry on with the Knight's Move (2 cells up & 1 cell to the Right) & with Breakmoves (moving 1 cell down at 10, 15, & 20) until the last cell is visited.

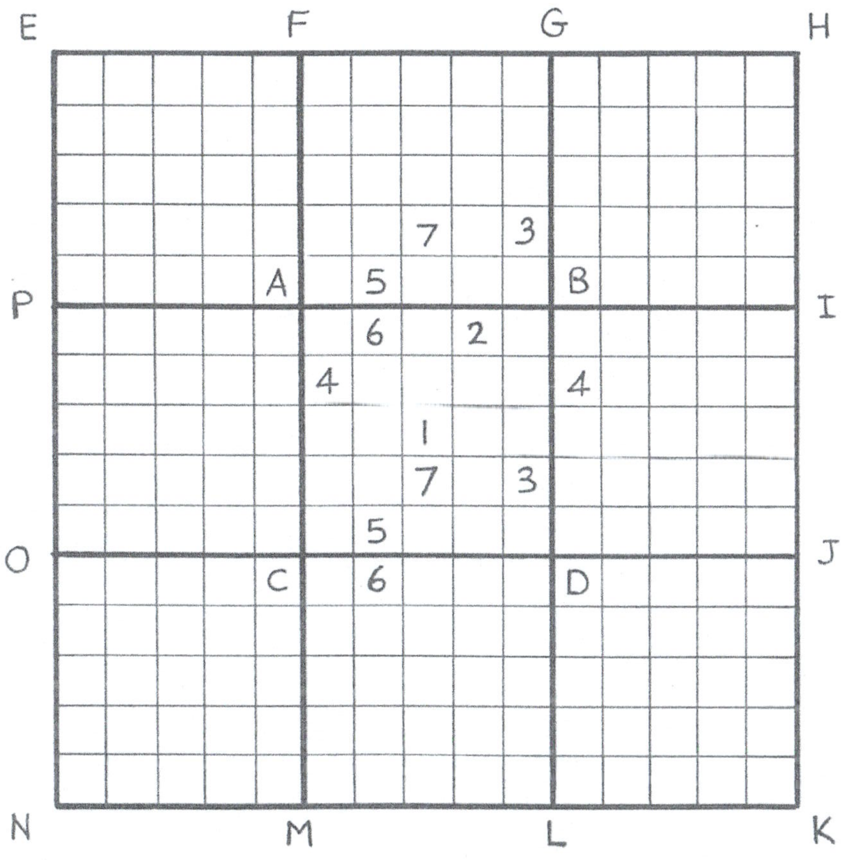

Fig 52

Page 74

So, the root cause is in the Mind it is in your interpretation. Buddha says, "Think that the whole earth is your friend". Why? Jesus says, "Even forgive your enemies". Not only that, "Even love your enemies". Why? Buddha & Jesus are trying to change your interpretations. But Tilopa still goes higher. He says even if you think that all are your friends, you continue to think in terms of friendship & enmity. Even if you love the enemy you think that he is the enemy. You love because Jesus has said. Of course, you will be in a better situation than an ordinary man who hates the enemy, less anger will happen to you. But Tilopa says to think that someone is an enemy, to think that someone is a friend is to divide – you are already fallen into the trap. NOBODY IS A FRIEND AND NOBODY IS AN ENEMY. THIS IS THE HIGHEST TEACHING...

OSHO

(excerpt from "Tantra" by Bhagwan Shree Rajneesh p 223 of the chapter "Cut The Root" Discourses on the Tantric Way of Tilopa's Song of Mahamudra).

UNI-CENTRED

Sufi Majik Sq. of 7×7.

13th Century Persian manuscripts gave great meaning to the fact that the final breakmove at the last number leads back to the No. 1 which begins another cycle of the Endless Becoming endlesslybecoming endlesslybecoming endlesslybecoming.

Thoughtful Muslims had long considered the Magick Squares as symbolic models of the Universe & now they could view them as representations of Life in Endless flux, constantly being renewed by a source of energy at the heart of the Universal Plan, Plane, Planet. Occasional manuscripts (In the Yale Library, and elsewhere) show Empty-Centered Squares, in which devout Muslims, apparently impressed by this, refused to set down the awesome middle number (the Holy Number One).

For certain artistic reasons, The Uni-Centered Sufi Sq. of 9×9 cannot be given. [whoever can create it will receive a GIFT, if they write to me]. Instead the Sufi Sq. of 7×7 will be constructed by you, on this page. Refer to the Solutions chapter at the end of this book.

			1			

FIG 53

To construct the UNI-CENTERED m. Sq. of 7×7 allow all moves to progress like the Knight i.e: 2 cells up & one across to the right. L-shaped moves. At every multiple of 7 there is a Breakmove which progresses one cell down.

Extended or ELONGATED Knight's Moves.

FIG 54

80	58	45	23	1	69	47	34	12
9	68	46	33	11	79	57	44	22
10	78	56	43	21	8	67	54	32
20	7	66	53	31	18	77	55	42
30	17	76	63	41	19	6	65	52
40	27	5	64	51	29	16	75	62
50	28	15	74	61	39	26	4	72
60	38	25	3	71	49	36	14	73
70	48	35	13	81	59	37	24	2

This Majik Square of 9x9 (Fig 54) is constructed by elongated Knight's Moves : i.e 4 cells to the right & 1 cell upwards with Breakmoves of 1 cell downwards (at every multiple of 9). All columns, rows & diagonals have an equal sum of 369. This method for constructing majik squares or harmony squares is ancient. The method of transforming these Numbers into Graphics is "original" to Jain. Nothing is ever really "original" or "discovered" only "RE-discovered". At some time in our Earth's history, its all been done bef-ore. Keep joining long continuous lines from beginning to end, prayerfully.

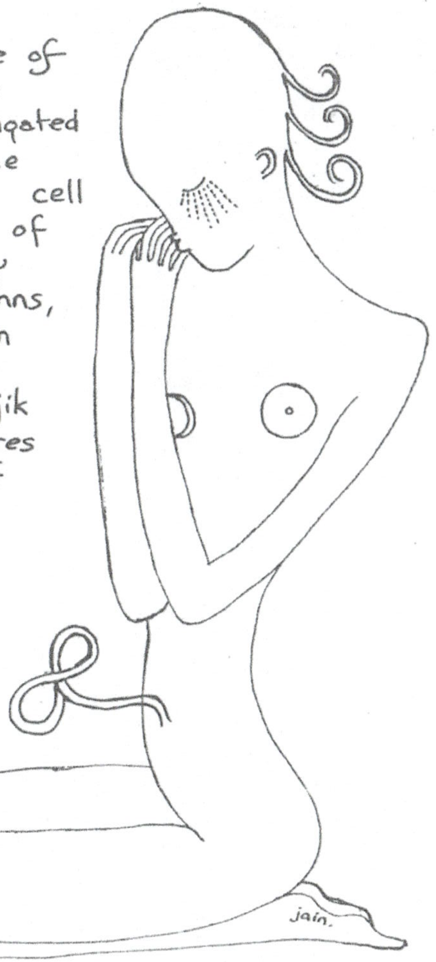

In the space of Fig 55 provided draw the 9x9 Majik Square Yantram at 0° of the Elongated Knight's Move presented in Fig 54.

Contemplate the word "YANTRAM" as not only meaning a "mystical diagram" but "as a machine or instrument for moving Consciousness".

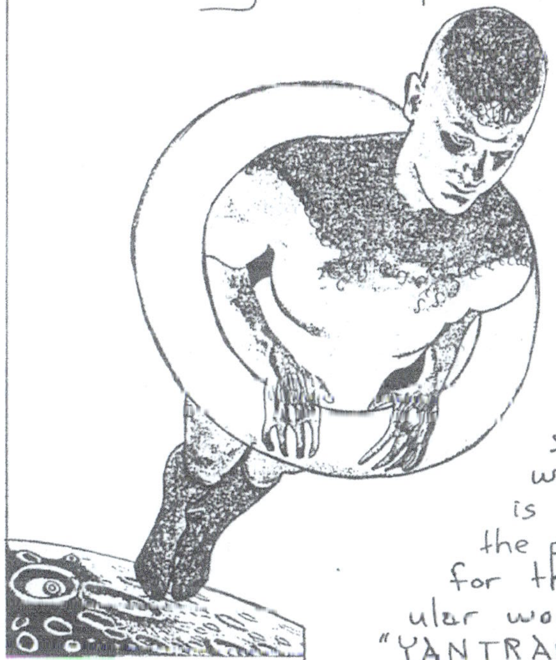

A technical spelling correction needs to be made here. The word "Yantra" is a sanskrit word but is actually the plural for the singular word "YANTRAM". Thus Fig 55 is a Yantram.

Page 77

This Yantram is dedicated to the Custodians of Light in the New Age

Fig 55

An ASSOCIATED M. SQ OF 9×9

47	58	69	80	1	12	23	34	45
57	68	79	9	11	22	33	44	46
67	78	8	10	21	32	43	54	56
77	7	18	20	31	42	53	55	66
6	17	19	30	41	52	63	65	76
16	27	29	40	51	62	64	75	5
26	28	39	50	61	72	74	4	15
36	38	49	60	71	73	3	14	25
37	48	59	70	81	2	13	24	35

Fig 56

As explained for Fig 47, when the Balance of The Pairs are equidistant from The Centre, this is termed "Associated". In Fig 56 :

- There are 20 different summations of 369.
- Notice how the centre column has all its digits ending in "1".
- On a separate piece of paper practise constructing this 9x9 Associated M.Sq.. Begin with "1" in the middle of the top row & continue all moves One cell up & One across to the right. Remember, as was explained in Fig 52, that at every multiple of 9 there is a BREAKMOVE (or Interruption) where that number is moved one cell DOWN.

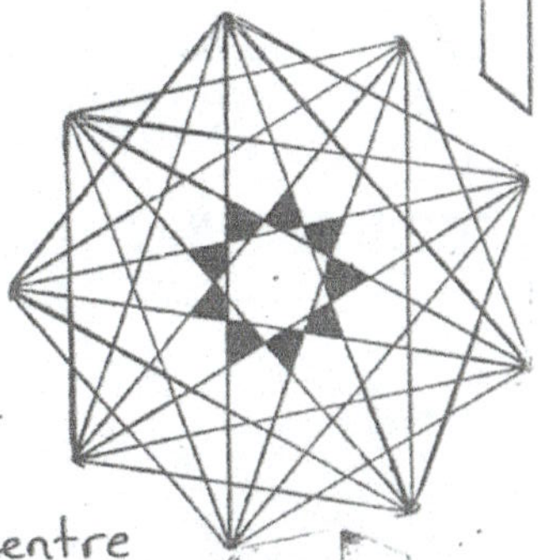

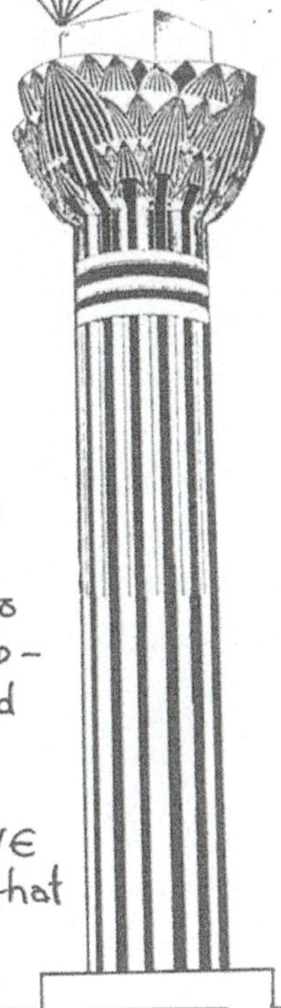

Carefully construct the Associated 9x9 majik Square Yantram in the space of Fig 57 provided?

Page 79

Fig 57

Creative Resolutions

Whilst creating the Majik Square Art in this Dossier of Sacred Geometry, have in your Consciousness, your Higher Self seated in the Throne of your Mind. Before & during the construction of a Majik Square Yantram put forth any resolves, any affirmations or any tensions to be dealt with. Treat these exercises with "Yantram Creations" as a time to go inward & come out, when completed, with Full Knowingness & Liberation. You are The Harmony that you have just created, with no beginning nor end; you are that long ancient continuous line spinning the pulsating mandala web in the playground of your Creation. Always, in the construction of these webs, let your breathing be gentle & rhythmic, keep your posture erect & your work neat. It is not just an academic exercise, it is a Yoga too ("Union"). These little designs will one day be muralized on the Peace walls of your Town, & the techniques that you have grasped will lead you towards a new Spectrum of Creative Works.

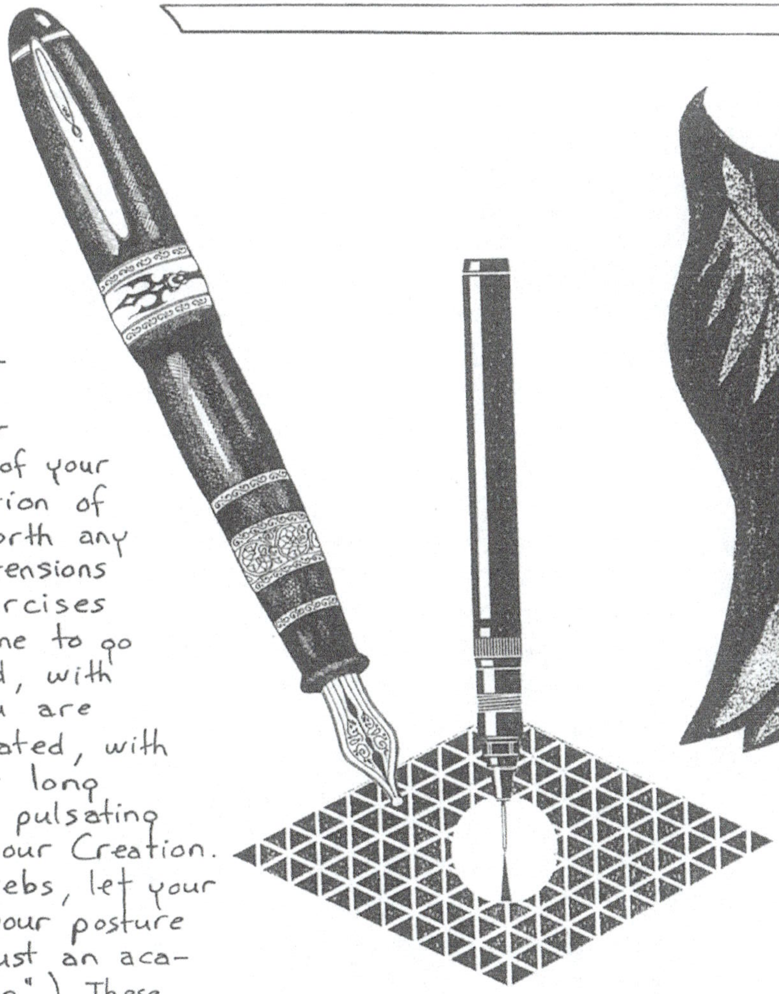

M. Sq of 9 Lozenge

An alternative method of constructing M.Sqs is by the Lozenge Method in which all of the odd Nos. (EFGH) are arranged sequentially in the form of a Square, the points of which meet the centres of the sides of the main square ABCD (fig 58) & the Even Nos, (Square GHIJ) fitting in the corners as shown in Fig 59. Notice how the Even Nos. are written also in a consecutive order of progression. →

Pick up Δ HIC & place it into Δ FGD.
Pick up Δ ICJ & place it into Δ BEF.
Pick up Δ GCJ & place it into Δ EAH, to form a main square that is magic, the result of which is shown in Fig 59.

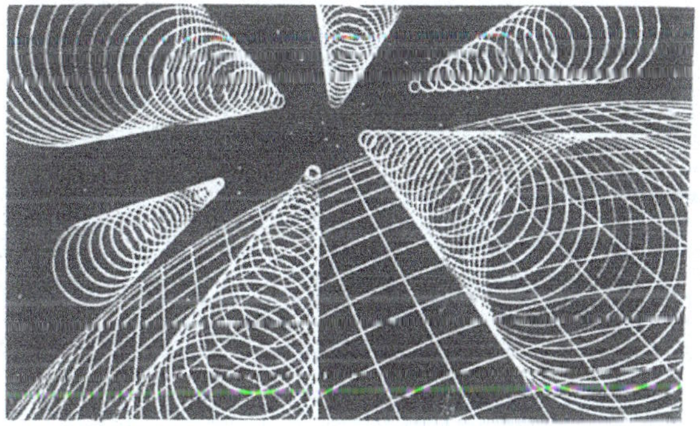

All complementary pairs are balanced about the centre. Thus Fig 59 can be defined as an Associated Lozenge Majik Square of 9x9 whose Constant is 369.

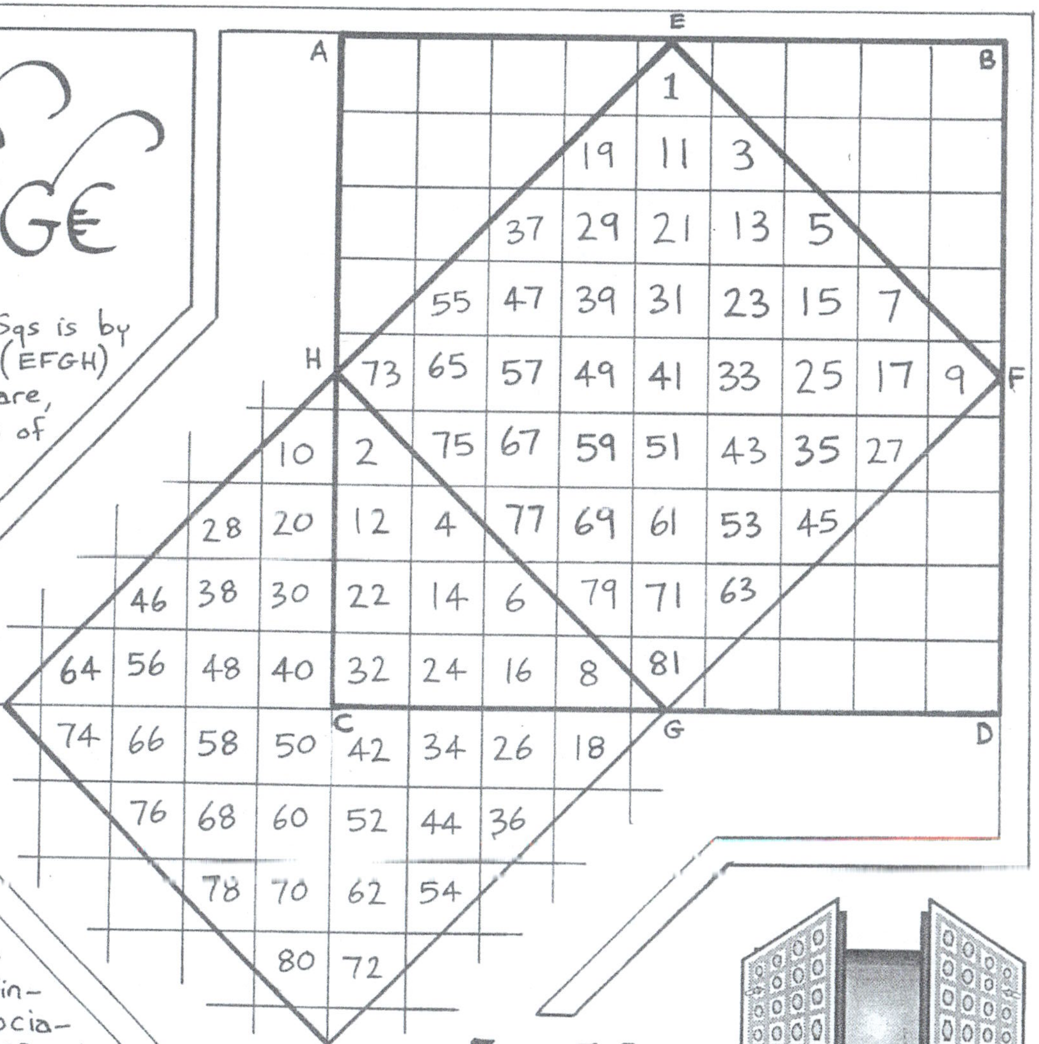

Fig 58

42	34	26	18	1	74	66	58	50
52	44	36	19	11	3	76	68	60
62	54	37	29	21	13	5	78	70
72	55	47	39	31	23	15	7	80
73	65	57	49	41	33	25	17	9
2	75	67	59	51	43	35	27	10
12	4	77	69	61	53	45	28	20
22	14	6	79	71	63	46	38	30
32	24	16	8	81	64	56	48	40

Fig 59

Page 82

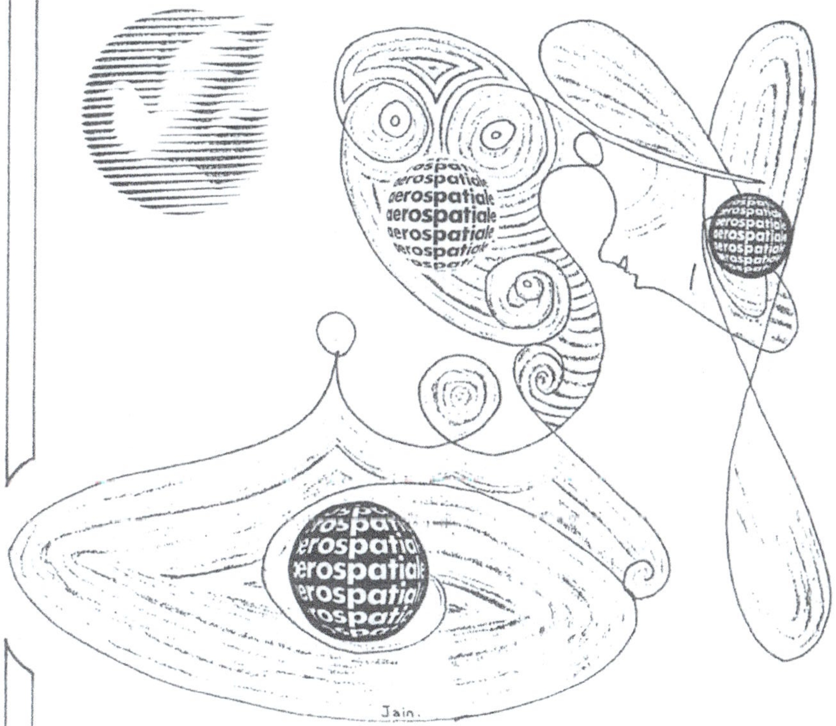

Upon the 9x9 Lattice or Grid of Dots provided in the Space of Fig 60 construct the Majik Square Yantram of the Lozenge Square shown in Fig 59.

Fig 60

Page 83

Composite Chinese M.Sq 9×9.

The most spectacular development of the Lo-Shu (3 × 3) pattern appears in the old Chinese Majik Sq. of Nine. To begin with, they first wrote the sequence 1 to 81 in nine parallel rows to form a 9×9 Natural Square as shown in Fig 61. (NB: The Diagonals of any Natural Square determines The Constant; both Diagonals of any Natural Square are always of the same sum & this determines what the sum of the columns etc of a M. Sq. will add up to).

6	1	8
7	5	3
2	9	4

Then they took each vertical column of nine to form a secondary Majik Sq. of the third order i.e 3×3. Finally, taking as a guide the lowest number of each small square, they set out the following M.Sq of 9×9 in the pattern of The Lo-Shu to make a single huge one. This is shown in Fig 62. Notice how the 9 underlined Numbers form The Lo-Shu.

1	2	3	4	5	6	7	8	9
10	11	12	13	14	15	16	17	18
19	20	21	22	23	24	25	26	27
28	29	30	31	32	33	34	35	36
37	38	39	40	41	42	43	44	45
46	47	48	49	50	51	52	53	54
55	56	57	58	59	60	61	62	63
64	65	66	67	68	69	70	71	72
73	74	75	76	77	78	79	80	81

FIG 61

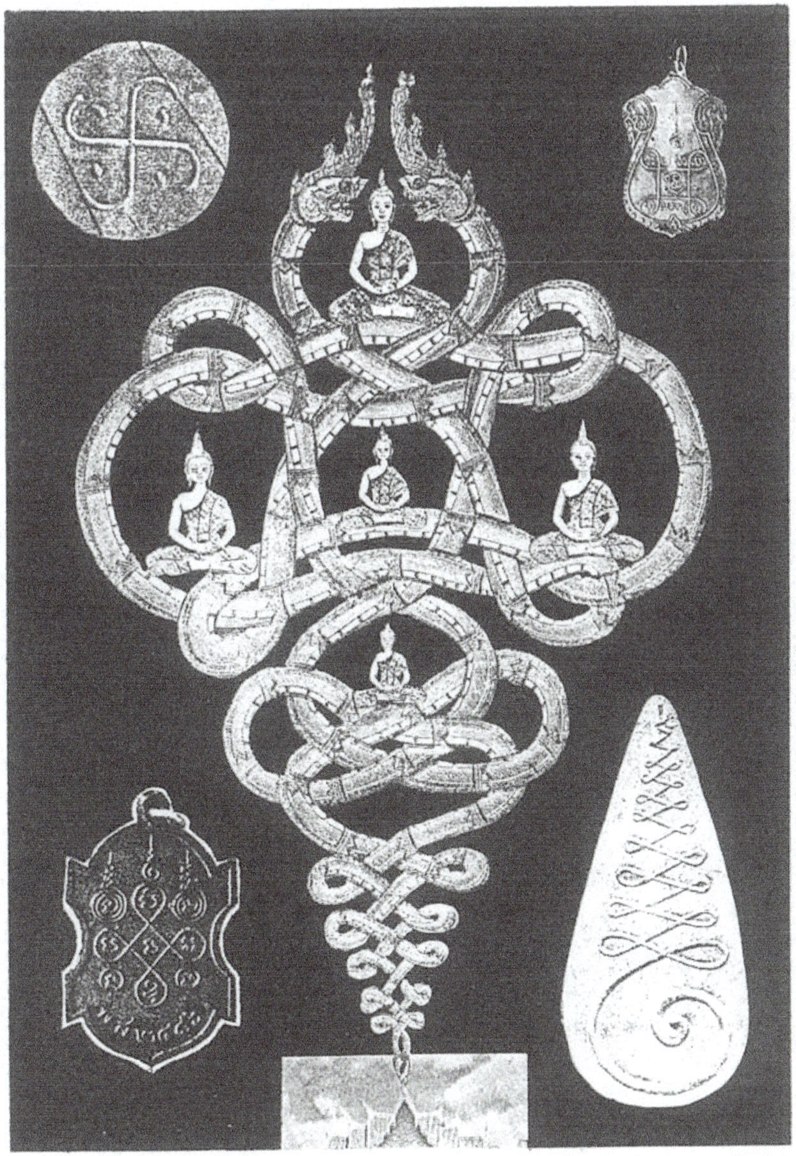

51	6	69	46	1	64	53	8	71
60	42	24	55	37	19	62	44	26
15	78	33	10	73	28	17	80	35
52	7	70	50	5	68	48	3	66
61	43	25	59	41	23	57	39	21
16	79	34	14	77	32	12	75	30
47	2	65	54	9	72	49	4	67
56	38	20	63	45	27	58	40	22
11	74	29	18	81	36	13	76	31

FIG 62

"The World is built upon the Power of Numbers" (Pythagoras, C6th B.C.).

The graphical yantram of Fig 62 is not necessary to be drawn; (The following m.Sq: Fig 63 will be drawn) but rather Fig 62 is to illustrate how a knowledge of the smaller Majik Sqs. e.g. 3×3 & 4×4 etc can assist in building up the larger ones, eg 9×9, 12×12, 16×16 etc.

The Balance of The Pairs shows that this is an Associated m.Sq. of 9×9 i.e: the complementary pairs or Diametrically Opposite numbers equidistant from the centre, have a sum of 1 + 81 ie: 82. Notice also that the sum of each pair is twice the middle number (41).

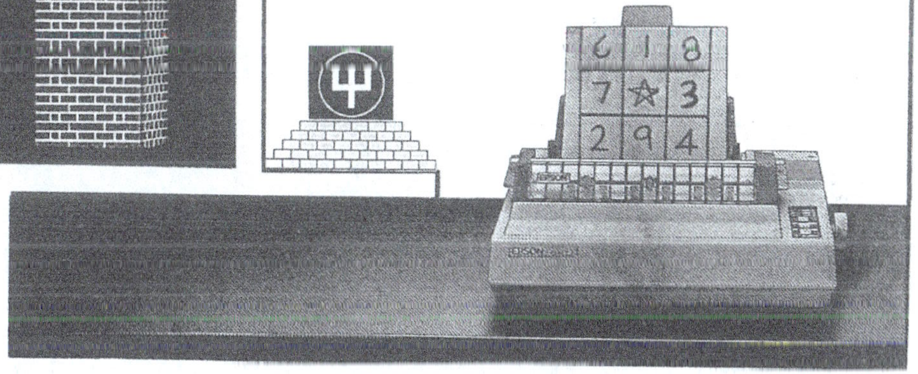

Page 85

ANOTHER COMPOSITE M.SQ of 9×9 based on the LO-SHU

- Each 3×3 sub-square is a m. sq.
- Each cluster of 9 Nos. in natural counting order moves to its next cluster of 9 numbers in the order of the Lo-Shu.
- The Sum of the 9 central cells of the nine 3×3 sub-cells also adds up to 369 !

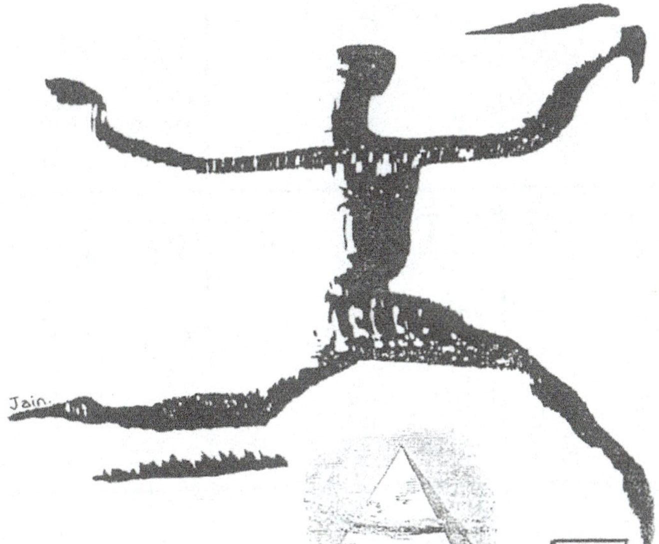

6	1	8
7	☆	3
2	9	4

51	46	53	6	1	8	69	64	71
52	50	48	7	5	3	70	68	66
47	54	49	2	9	4	65	72	67
60	55	62	42	37	44	24	19	26
61	59	57	43	41	39	25	23	21
56	63	58	38	45	40	20	27	22
15	10	17	78	73	80	33	28	35
16	14	12	79	77	75	34	32	30
11	18	13	74	81	76	29	36	31

Fig 63

DIAMONDS OF WISDOM

Fig 64 CRYSTALIGHT

Page 87

THE GIANT PYRAMID(S) OF SHENSI.

You may wonder why this page of a Chinese M. Sq is adorned with Egyptian symbols. It is to inform the deceived public that 16 pyramids exist today in central China, one that is twice the size of the Egyptian pyramid at Gizeh. These 16 pyramids are very similar to the ones scattered along the banks of the Nile in Egypt, acting like electrical stations, geometrically in phase so as to pull into the Earth an infinite "Free Energy System" 'cos the Earth acts like a giant transmitter. The pyramids on Mars are also linked to our planet. (The Arabic word "EL KAHIRE" was the ancient name for "MARS" & is today called "Cairo".

All pyramids & sacred sites lie on the Earth's auric/electromagnetic Ley-Lines which circle the Earth with numerous intersection points.

Looking at Fig 63, construct the M. Sq. Yantram at 0° in the space of fig 64.

Page 88

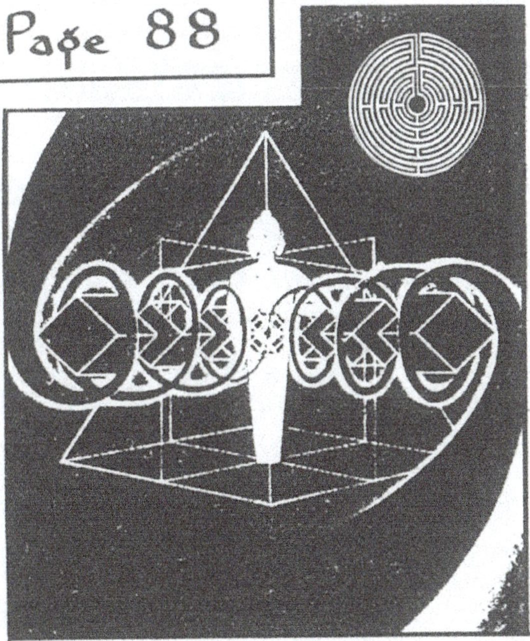

THE ancient Pyramids were temples for the living, not the dead. They are lined up on the magnetic grids of the earth. The triad on the face of the pyramid represents astral, mental & physical. Astral is the Spirit, mental is the conscious desire, & physical is the manifestation. When you bring them together, they form the energy of the fourth level.

(Ramtha)

from THE VIEW OVER ATLANTIS
by John Michell. 1969 Sago Press.

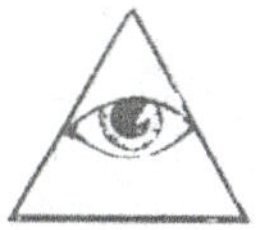

THE GREAT PYRAMID is, as we have seen, a model of universal growth. Its measurements, springing from the invisible mote at its very tip to expand in stone through the human scale of perception and pass beyond to encompass the earth. Its height, a ten millionth part of the distance between the Earth and the Sun, demonstrates the unity of terrestrial and heavenly geometry. Certainly the whole edifice was built by men who knew something of the laws which govern the intergalactic flow of living energy and of their application to the development of life on Earth. By acute observation of nature, leading to the detection of certain numerical patterns behind the process of growth, it is possible to gain complete knowledge of the universal system, verifiable both by reason and directly by the senses.

The perfect geometric figures, to which imperfect natural forms relate, develop from the basic structure of the cell, and this structure is shaped by particular combinations of the elements. Thousands of years before Kepler explained the mechanics of the universe through the manipulation of the five perfect solids, a system of three-dimensional geometry had been developed to illustrate every possible form of growth. The chief cosmic intervals and numbers, together with the ratios that determine the patterns of life, were related to each other by means of numerological combinations such as magic squares. Within the crystalline structures of figures was stored the universal scheme.

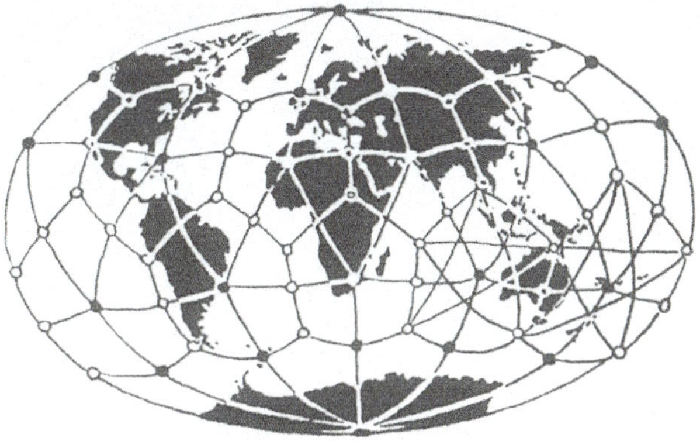

PLANET EARTH'S CRYSTAL GRID

It
is time
to light the lanterns again
to see the imageless God that you are.
imageless
IMAGELESS
IMAGELESS
I'mAGELESS
I'm AGELESS

THINK GLOBALLY, ACT LOCALLY

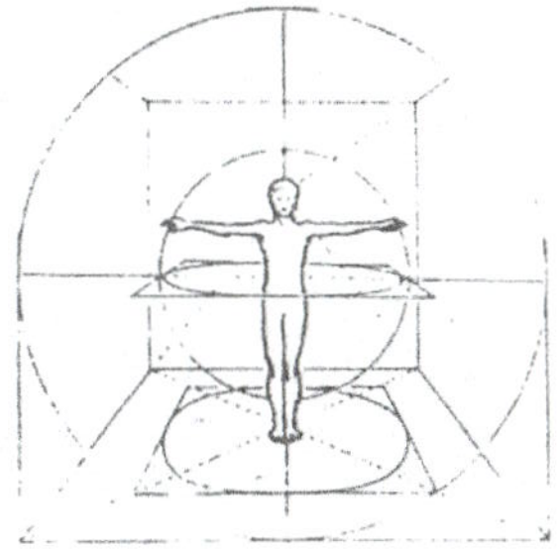

Go on your Way
& breathe in the colors of the Sun
the Earth, the Skies & the Seas
& know that what is Above
is Below

Wear the Elements about you
like a temple Robe
& know that Spirit is your Passion
of your Art that makes time Timeless

Let Nature be your Friend
& Sounds of Birds be your Song,
Meditate with the Wind
for it has much to tell you
there is much for you to know
to Open Heart Surge

Go serenely my Friend
for the Majik is withIN you,
Express the ripples of your Unfolding Star
Play your Majik Rainbow 7even-Stringed Guitar

Know that I Love you with great depth
& if ever you need
to pour your Soul into the Stars
just create It in your Mind
'cos there is StarDust in our DNAs
& Golden C(h)ords between us
so beautyfull to have a Friend.

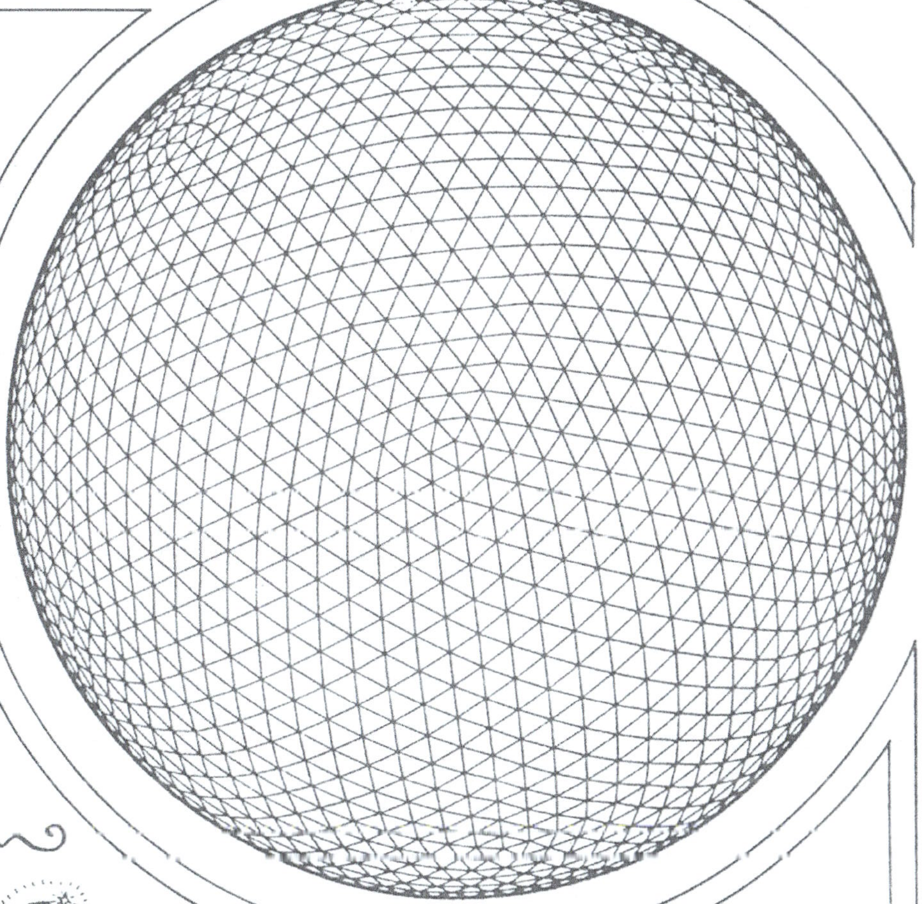

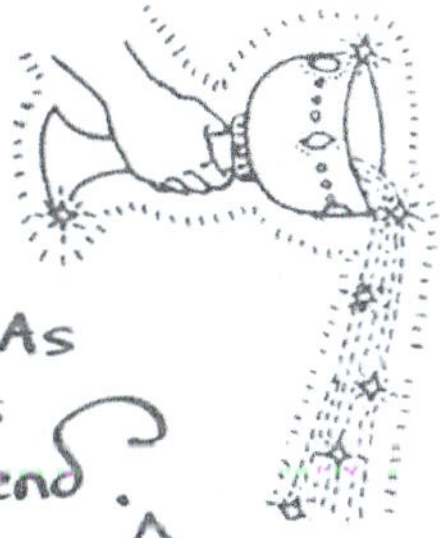

(Prayer by Jain).

Page 89

for the Construction of the

ENNEAGON.

(Nine – Sided Polygon).

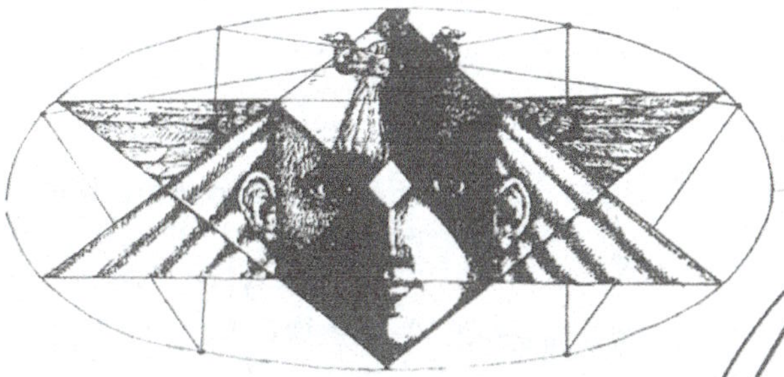

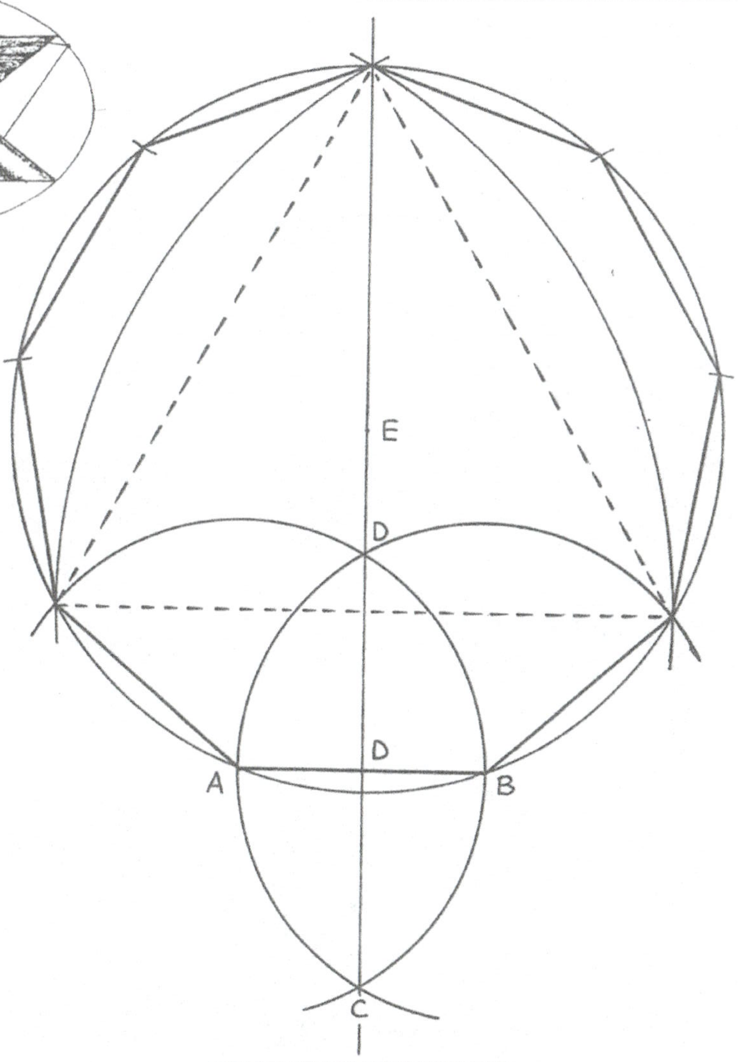

Fig 65

In the space of Fig 66 you will repeat the diagram of Fig 65. The desired initial edge of the Enneagon is given as the length AB. By drawing a fish shape (or VESICA PISCIS as it was called in the medieval times), that is an arc which is exactly one third of a full circle centred first on "A" & then repeated with "B" as centre, & extending these arcs to become exactly half a circle on each side, we then take the centre line from the lower intersection "C" to the upper "D" & extend this upwards. If we halve the distance between "A" & "B" & measure this **distance** from point "D", going up the centre line, we can mark point "E" which will be the centre of the 9-side Polygon. Place your compass on point "E", & with Radius "BE" inscribe the circle of which the initial line "AB" will be one-ninth. It is now possible to mark the 9-fold division of the Circle, with a fair degree of accuracy

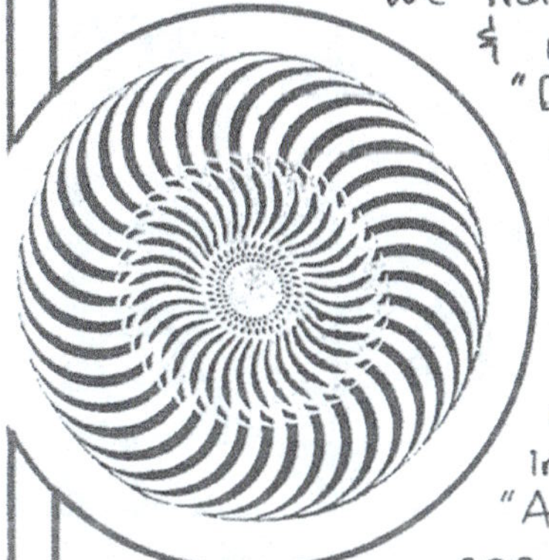

Page 91

Fig 66

Introducing the metaphysically aesthetic "SHRI CHAKRA", also known as: "SHRI YANTRA"

CHINNAMASTA

छिन्नमस्ता यन्त्र

Fig 69.

Before describing the Sri Chakra (Yantra of 9 interlacing triΔngles), I will give a few pages of notes on the definition of a "Yantra". Yantras can be very simple symbolic graphical statements of our universe. In Fig 66 you constructed an exact Enneagon. Thus Fig 67 shows an Enneagram: the relationship of nine points around the circumference of a circle. Each point corresponds to a bodily aperture eg, eyes, ears, anus.

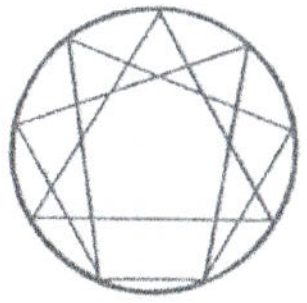

Fig 67

In 1990 in Ajmer, India, I met an extremely youthful 80 years young Yogi with long blond hair & beard to his waist/sarong. He often talked about "Tantra, Yantra, Mantra, Jantra" (which is pronounced "Tuntra, Yuntra, Muntra, Juntra"). His name is BAPUJI & scientists around the world have monitored his heart & breath, for he is famous for being able to remain buried underground in a coffin for 2 to 3 Days! (i.e ability to slow down his heart beat etc). He wore a Juntra which is a M.Sq. imprinted on copper about 10 x 20 mm & worn on the biceps as a sign of spiritual strength. See Fig 68. This M.Sq. amulet worn for protection has summations of 20. Thus this is called a Bisa Yantra. ("Bisa" means 20). I spent a few days with Bapuji & his meagre room consisted of framed Majik Squares on the walls & books on M.Sqs in Sanskrit, some of which are now in my possession.

The inverted triangle of Fig 69 always represents the Vagina or Yoni.

Fig 70 shows a Vedic fire altar in the form of a magical triangle, associated with rites for gaining strength & supremacy. Such Yantras used for ritual worship have evolved into altar-plans which express the mathematics of the cosmos.

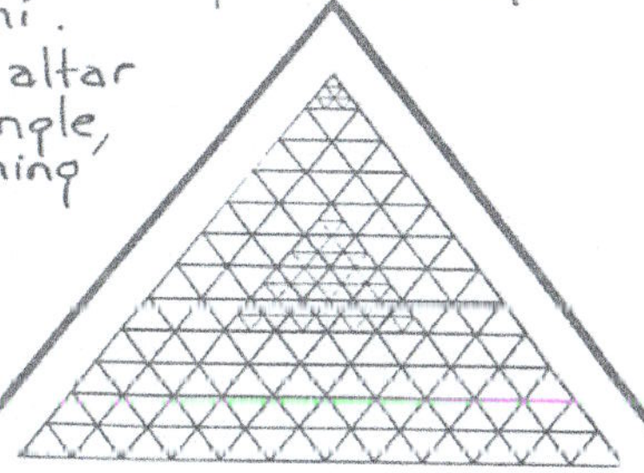

Fig 70

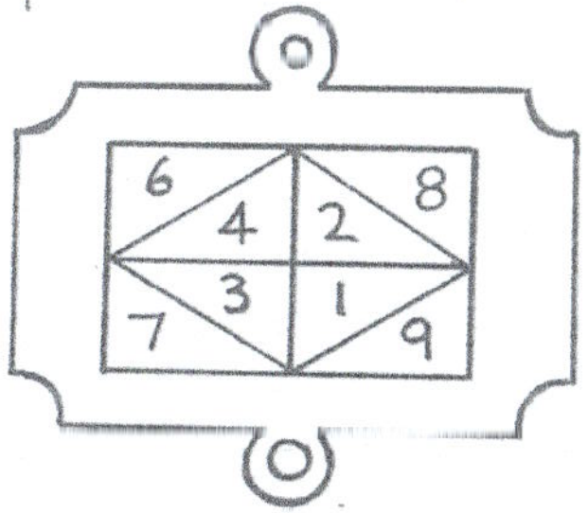

Fig 68

Page 93

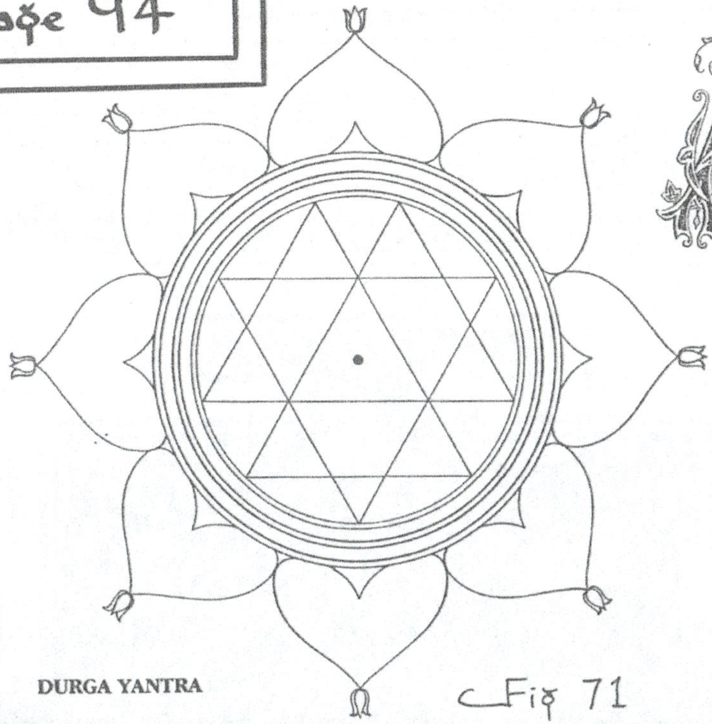

DURGA YANTRA

Fig 71

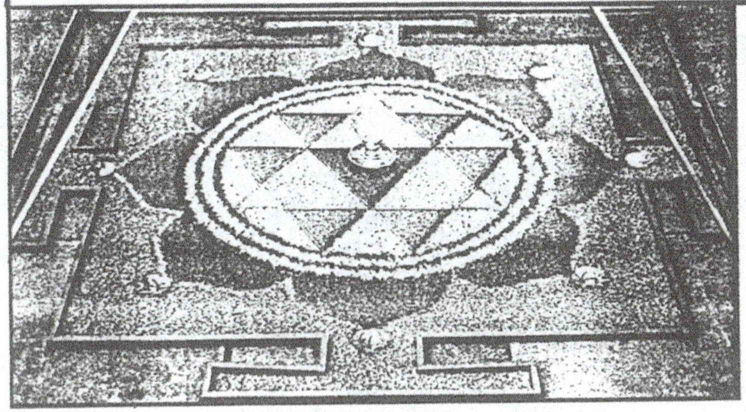

tantrik (adept of tantra), with the aid of various Yantras can exercise influence on the Deities, natural forces or phenomena. By means of yogic practices, they can unite the internal Siva & Sakti & elevate the human soul to the status of ASMI "I Am, GOD I AM". This process of Self Realization is the main aim of Tantra. Yantra is a tool. Yantra is a diagram, geometric symbols of the Deity. Durga Yantra (Fig 71) is actually The Abode of that particular Goddess; (Durga is another aspect of Shiva's consort Shakti). A Deity is contacted by these Yantras. Every Deity has a mantra (a magical sound formula or oral spell) whose syllables can determine the shape of the Yantra. Of course, the followers of tantra do not give out or divulge these secrets to persons outside their fold. Hence, it is difficult to know the true nature of the mystery behind the Nos. as conceived by the tantriks. But the tantriks did have a clear idea of the famous "majik squares" from mathematics, though they never explicitly expressed it.

Any 2-Dimensional Yantra is meant to be meditated upon as a 3-D perfect hologram, a geometrical paradigm, a mystical construct of the universe in which all polarities are harmoniously united. A Yantric artist must control his/her impulses, fast, pray: "O thou Lord/Lady God within, teach me in dreams to carry out the work I have in mind..." Priests of Inner Poise consecrate the Yantra & endows it with sared Power, with mantra, with clear focussed thoughts. He/she must feel the very nature of the energy to be transmitted to the Yantra affectinG all who view It.

IN INDIA

A BANQUET FOR ANTS, consisting of wet rice flour, is spread out in a complex design before a temple. In South India, Hindu women create these designs in front of their homes on holidays as good luck symbols.

The yantra-siddhāis (sages to whom the powers of the yantra have been revealed) declare that if a yantra fails to become 'psycho-active' the fault lies with the person using the yantra & never with yantra itself, for such power diagrams that move consciousness are based on the laws of space, vibration, number & dynamics. The maker of deformed yantra is subject to psychic disequilibrium. The Yantras' rhythmic lines of the laws of harmony referred to an ancient Vedic principle called "Ardhamatra" which is the mathematical descent from the Bindu (Centerpoint, sperm, drop) to the unified & ordered periphery ('rdh' = to grow, & 'matra" = measure, in sanskrit language).

Shown in Fig 72 is a 1761 Nepalese Yantra 'The Nava-Yoni Chakra' floating in the cosmic waters. This signifies the creation of the Universe by the union of the male & female principles, represented by interpenetrating upward & downward pointing triangles. The 9 inscribed triΔngles indicate the 9 (nava) cosmic wombs (yoni). Such an icon derives its power from the person constructing it. It is the cultivated intentionality of the devotee (as well as the traditional design) that makes the yantra function as a powerful facility. This process of transmitting love power into a design is technically known in ancient Bharat (ie: the original name of India) as "MANDALA" (i.e: "that which gathers the essential details", mandam lāti). In a state of Concentration, the centre of the devotee's own being is The Mandala. It is a psycho-cosmic mechanism for effectively relating the inner space. Thus these psychocosmograms are models for transformation, projection, concentration, integration, dissolution & absorption. They represent the universe as an enclosed field for the play of consciousness. The entire mandala (pron: "mun-dala") is viewed as a city, palace or island which is the spiritual residence of the Hindu Gods. The human body is also a mythical Mount Meru (resplendent abode of the Gods) Mandala with its highest peak in the thousand-petalled lotus in the brain where the Self resides. This chakra (or energy centre, lit: "wheel") is called Sahasrāra). Different chakras are equated to various planetary energies (Fig 73) & are a rich inspiration for mandala construction.

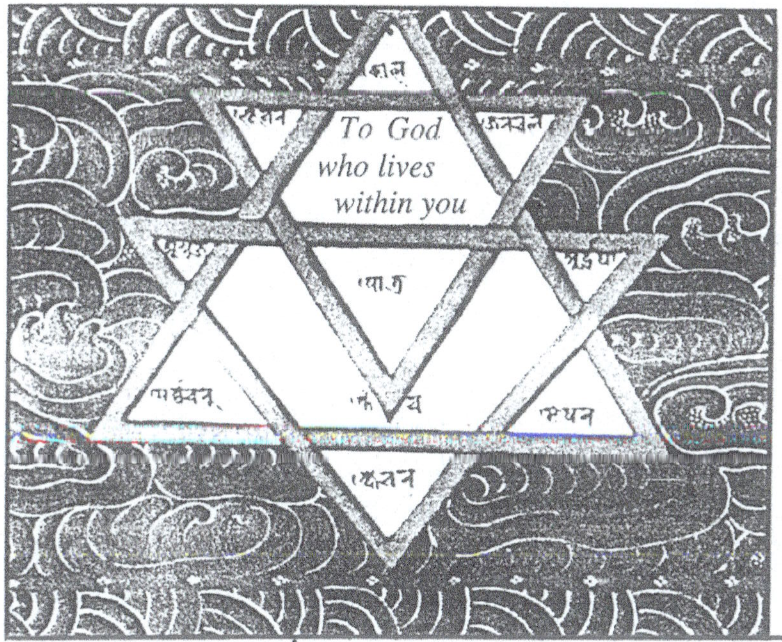

Fig 72 *Indian Heritage*

9	4	11
10	8	6
5	12	7

Mercury

11	6	13
12	10	8
7	14	9

Venus

7	2	9
8	6	4
3	10	5

Moon

10	[illegible]	[illegible]
11	9	7
6	13	8

Jupiter

[illegible]	[illegible]	[illegible]
7	5	3
2	9	4

Sun

[illegible]	[illegible]	[illegible]
9	7	5
4	11	6

Mars

13	8	15
14	12	10
9	16	11

Rahu (North Node of the Moon)

12	7	14
13	11	9
8	15	10

Saturn

14	9	16
15	13	11
10	17	12

Ketu (South Node of the Moon)

NUMERICAL YANTRAS

Fig 73

Fig 74

SHRI YANTRA

The cult of Sri Chakra has spawned several temples of worship, especially in South India. According to the ancient Vedic corpus, Sri Chakra is essentially worship of the mother-goddess called variously, Lalita, Shodasi, Tripura-Sundari, Adyā Nitya, Dévi. They are invoked as the central divinity in this symbol of fecundity (yoni or vagina is symbolized by the basic form of the triangles). The 'sexual' aspect of Siva-Sakti is not to be viewed as a divine coupling but rather the unfolding of cosmic poles producing two forces, matter & anti-matter. It is the spark of creativity or Life arising from the union of male & female principles, in which this world of chaos & multiplicity is held by The UNITY & Harmony of the primordial Bindu (or centrepoint). Each triangle (of which there are 43) of this yantra has its presiding deity or divinity representing the subdivided energy-self of the Great Goddess. The true purpose of the Sri Chakra was to develop "inner worship" not to perceive it as an erotic symbol.

Anyone undertaking to prepare the Sri Chakra (Fig 74) must be proficient in the Sri-Vidya (Wisdom) lore. It must be prepared as an act of worship, with purity of body & mind, fasting & in seclusion. After preparation he/she must use it on their own altar & then pass it on, not accepting money or gifts. It's efficacy consists in its being breathed into, touched & given by a holy man; it has no value when bought in a market.

Fig 75

The CONSTRUCTION of the SRI CHAKRA is a unique & complete rite. There are two main methods of construction: one used by the "right-hand' path tantrikas, is the order of evolution (sristi-krama) in which the yantra is constructed in geometrical progression from bindu outwards. This is shown in Fig 75.

The other, the order used by the 'left-hand' path tantrikas or yogins, is the order of dissolution (laya-krama) in which construction is from the circle moving inwards. The 9 parallel & equidistant lines within the circle constitute the Sri Chakra. These nine lines represent the nine stellar bodies which regulate the activities on earth. they are sun, moon, angāraka, budha, brhaspati, sukra, sani, rāhu & ketu. The 4 triangles pointing upwards are called Siva-Chakra (the male aspect) & the 5 triangles pointing downwards are called Sakti-Chakra (the female aspect). The central point is thus the Union of the ultimate male principle (Purusha) & female (Prakrti). In tantric ideology this chakra shows the Goddess as dynamic. This yantra is enclosed by two concentric circles called lotuses, the inner one having 8 petals & the outer 16. 3 circles engirdle the petals then the whole design has an enclosure of triple lines called "bhupura". On each side of the square is a portal or gateway (dvāra) which means only the honest & devout can enter into the mysteries of Sri Chakra. The Yoginis (Unifers) are the powers that aid the union of the devotee with the Mother-Goddess.

Page 97

Page 98

A traditional rangoli pattern from Tamil Nadu made at the entrance of the house

NAMASKAAR

नमस्कार

THE MASTERS

DISCOVER THE AMAZING WORLD OF

MEDITATION

BECOMING

THE TEMPLE OF LIGHT

TO BE QUIET WITHIN

The form of Sri CHAKRA Yantra is prescribed in the Tantra Shastra. The significance of this yantra is that this Chakra consists of dots, straight lines & circles. It is well known that the property of matter, if geometrically represented, begins with a point & proceeds through straight lines or combinations of straight lines like triangles or quadrangles or in circles. Therefore it is the representation of the entire universe.

The proper way to worship Sri Cakra is as a 3-dimensional yantra of moving force, envisioning it with height, a gold, jewel-studded mountain/palace of nine tiers which is the home of The Goddess. Each tier is known as an Avarana. These nine Avaranas are together called Navavarana. The last tier is the Bindu, known as Sarvanandmaya Chakra. In this Bindu peak/crown the Maha (Great) Sakti has taken her abode. Devotional hymns are sung to Maha Sakti. When you meditate on this Shakti Chakra, the love of the Universe is sublimely drawn into you ∞.

Fig 76

The worship or cult of Sri Chakra presented in this sub-chapter of M. Sqs. of 9 is but a glimpse of the complete esoteric picture. Every line & every shape of this yantra has its Keeper of the Secrets who stand like guards e.g: the outer-most line surrounding the square (of Figs 74 & 76) which is the residence of the 10 spirits' accomplishment (siddhis), four of them stationed at the 4 portals, another four at the 4 corners (called upa-āmnāyas) one above (urdhva) & one below (adhara). These spirits represent abilities to become minute (animā), swell into magnitude (mahimā), to become light (laghimā), to become heavy (parimā), to lord over (īsitva), to desire (icchā), to subjugate (vositva), to attain the objects of desire (prakāmya), to get pleasures (bhukti), & to actualize all intentions (sarva-kāmasiddhi). These are powers to be acquired for self-protection along the spiritual journey.

Such detail can be given about the divinities that occupy the 16 petals etc.

Fig 76 is thus considered the most important of all Tantrik yantras (C 17th Nepal).

Yantra

THE TANTRIC SYMBOL OF COSMIC UNITY

Page 99

Fig 77

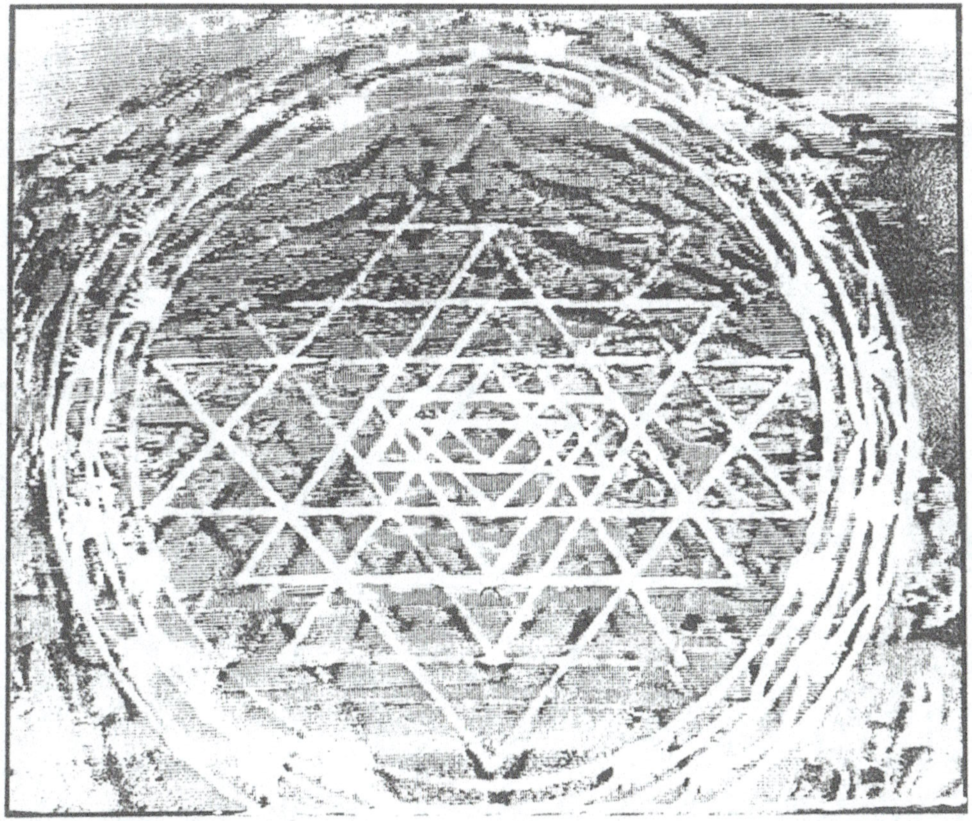

This rare photo shown in Fig 77 is a still from a film by Ronald Nameth. For the first time, Science/Technology has captured the metaphysical aesthetics of Yantric Art. Here, the Sri Yantra has been CREATED IN AN ELECTRONIC VIBRATION FIELD, an experiment in the translation of sound into vision. A similar experience is 'sensed' during ritual worship when the yantra pattern 'dematerializes', appearing to dissolve into a sound-pattern or vibration field of spoken mantras.

Yantra

Alexei Kulaichen of Moscow State University recently produced an algorithm describing Sri Yantra, a mystic diagram created in ancient India. This accomplishment, requiring many years of careful research, enlivened among scientists a new-found respect for the voluminous yet complex wisdom of India's antiquity. Yantras are diagrams composed of spiritual tantric symbols, designed to convey special knowledge and power. Only recently has science begun to ponder their enlightened significance. For Mr. Kulaichen's work a computer was required—a tool not available yesteryear.

TOWARDS THE RESACRALIZATION OF THE SRI CHAKRA & M. SQUARES.

४७	५८	६९	८०	१	१२	२३	३४	४५
५७	६८	७९	९	११	२२	३३	४४	४६
६७	७८	८	१०	२१	३२	४३	५४	५६
७७	७	१८	२०	३१	४२	५३	५५	६६
६	१७	१९	३०	४१	५२	६३	६५	७६
१६	२७	२९	४०	५१	६२	६४	७५	५
२६	२८	३९	५०	६१	७२	७४	४	१५
३६	३८	४९	६०	७१	७३	३	१४	२५
३७	४८	५९	७०	८१	२	१३	२४	३५

Fig 78

Presented in Fig 78 is the same Majik Sq. of 9x9 shown in Fig 56 except it is written in Sanskrit.

This Chapter Nine now concludes with a prayer: that you easily assimilate this new information & are Nourished by it. In the Name of Ardhanagesh-wari (Siva – Sakti as One embodiment) I pray that you worship no icon, follow no god but question all things & all teachers. Quietly express & enjoy your Truth.

Page 102

Mathemagic

IN SEARCH OF PATTERNS
DIGITAL SUMS

Pattern Book

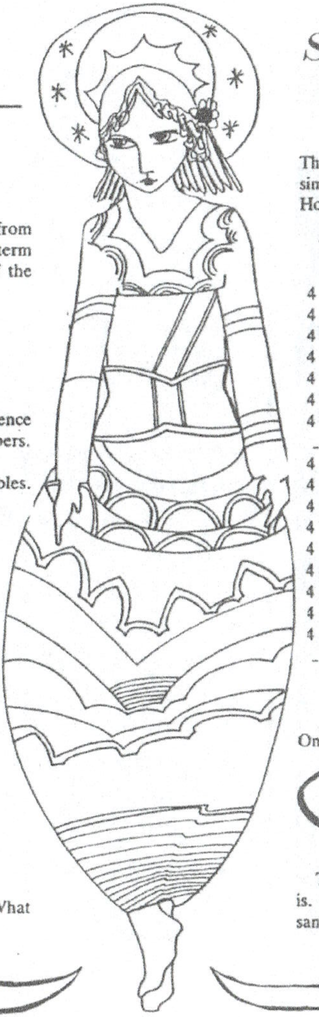

Mathematics is the study of patterns. In this unit we shall look at the patterns, generated from number relationships, formed when elements are reduced to their digital sums. Often the term reduced number is used and indicates the single digit obtained by continued addition of the integers.

For example, if the digits of 14 are added (1 + 4), the reduced number is 5.

Other examples: $324 \rightarrow 3 + 2 + 4 = 9$ $\qquad 226 \rightarrow 2 + 2 + 6 = 10 \rightarrow 1 + 0 = 1$

The number of steps in the exercise is called its persistence number. In 21315, the persistence number is 2 because $21315 \rightarrow 12 \rightarrow 3$. Find the persistence number of a selection of numbers.

Reduced numbers form repeating cycles called periods, some being as large as 24.

Smaller periods are obtained by finding the reduced numbers of the multiplication tables. Look carefully at the following:

		Product	*Reduced Number*
2 × 1	=	2	2
2 × 2	=	4	4
2 × 3	=	6	6
2 × 4	=	8	8
2 × 5	=	10	1
2 × 6	=	12	3
2 × 7	=	14	5
2 × 8	=	16	7
2 × 9	=	18	9
2 × 10	=	20	2
2 × 11	=	22	4
2 × 12	=	24	6
2 × 13	=	26	8

PERIOD OF 9

NEWS from the ART WORLD

The last digits in the *Product* column also form a repetitive sequence {2, 4, 6, 8, 0}. What would you expect the next set of reduced numbers to be?

When shown on a clockface, with digits 1 – 9 inclusive, this pattern emerges:

Simplicity

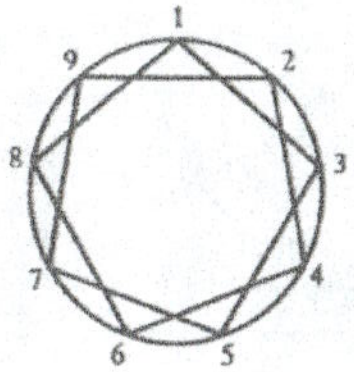

Fig 79

The three-times table has a much shorter set of repeating reduced numbers. Write them out in a similar way to the above. Find the period.
How are the elements connected?

The four-times table gives a more interesting set of results.

		Product	*Reduced Numbers*
4 × 1	=	4	4
4 × 2	=	8	8
4 × 3	=	12	3
4 × 4	=	16	7
4 × 5	=	20	2
4 × 6	=	24	6
4 × 7	=	28	1
4 × 8	=	32	5
4 × 9	=	36	9
4 × 10	=	40	4
4 × 11	=	44	8
4 × 12	=	48	3
4 × 13	=	52	7
4 × 14	=	56	2
4 × 15	=	60	6
4 × 16	=	64	1

The period of 9 is {5, 9, 4, 8, 3, 7, 2, 6, 1} *which is equivalent to the remainders when the product is divided by 9.*

By arranging the digits into two alternating columns two distinct patterns are formed:

5	
	9
4	
	8
3	
	7
2	
	6
1	

On the clockface:

Fig 80

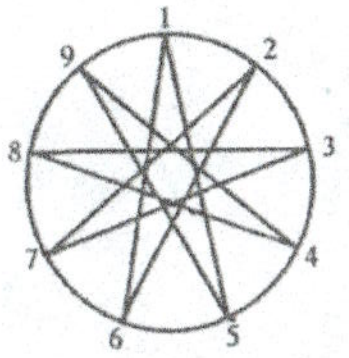

The five-times table is similar to the four-times. Write it out in full to see what the difference is. Go as far as 5×18. Alternate the reduced numbers. The clockface pattern should be the same, although it is done in reverse.

SONG of The MAJIK SQUARE MAN.

I know
the Majik Square Man
always painting
those vibrating number harmonics
modern yogi
from ancient Bharat[1]

I know
the Majik Square Puzzle Man
the Priest of Numbers
Theatre man of Holy Integers
in the Church of ancient
WoMandalaLand
Keeper of the Lost Scriptures &
Guardian of the sacred PHI &
Cubing the Sphere Geometries

Well I know
this Rainbow Warrior
Looking for patterns in the Sand
Atomic Art patterns in the Sand[2]
O my master Pleiadian
mandala geneticist

Well I savoir
this Farmer of Majik Squares
planting the precious Seeds
in the Hearts & Minds of our
Children
in Love
& Trust

The SONG of
THE LOVE OF DOLPHINS

O, majik Squares are the Blossoms
amongst the Weeds
Archetypes of Divine Order
in this fleeting world of Chaos
for it is the Gold extracted from the Dross

O Majik Square Man
WHEN YOU FLOW THRU me
& my Seven Daughters[3]
fill us with your Wisdom
fill us with your Vision
Overflow us deeply with the
Peace of your Inner Kingdom

When there is Confusion all about &
Life reeks with untruths & greed &
secret government satellites hypnotise you
that is when the M. Sq. Man Appears
The MATRIX AVATAR OF LOVE
designing symbols of Renewal
& Leading the Path of Non-Judgement

[FOOTNOTES: "1" = "Bharat" is the ancient name for India. "2" = The silicon molecule is proven scientifically to be an ornate ✡ mandala. "3" = The 7 Sisters are the 7 stars visible in the constellation of THE PLEIADES, the Jewel Box glittering in the night sky].

Page 103

INTRODUCING

CHAPTER TEN.

"OF TEN-NESS"

MEMO

FROM THE DESK OF

EARTH·MOTHER

For the Spirit
of the Majik
Square of Ten
to Appear in.

Sometimes known as
"The Magic One Hundred"
or "The Puzzle of Five
Hundred & Five".
For whoever has the Knowledge
: to arrange the first 100 Nos.
such that each Vertical,
Horizontal & Diagonal
Columns = 505
& the Sum of All = 5050

Page 105

Majik Sq. of TEN

You may wonder why only one M. Sq of 10 (fig 81) is given in this chapter. There exist millions of M. Sqs. of 10, but the few yantras that have so far been drawn do not produce symmetrical/aesthetic designs. Though, there may exist some M. Sqs. of 10 producing symmetrical / elegant designs. The M. Square Yantras, original to Jain, presented in Vols 1 & 2, are the cream of 18 years research. But who is to say that an asymmetric yantra is not aesthetic or of any use. So, in the space of Fig 82, construct the M. Sq Yantram at 0°, or if you want, you can just cheat & look at the result in the Solutions Chapter at the back of the book.

Scratchhh
Mmmm
Hahh
errr
Ohh
Ehh
ee
?

92	99	1	8	15	67	74	51	58	40
98	80	7	14	16	73	55	57	64	41
4	81	88	20	22	54	56	63	70	47
85	87	19	21	3	60	62	69	71	28
86	93	25	2	9	61	68	75	52	34
17	24	76	83	90	42	49	26	33	65
23	5	82	89	91	48	30	32	39	66
79	6	13	95	97	29	31	38	45	72
10	12	94	96	78	35	37	44	46	53
11	18	100	77	84	36	43	50	27	59

CREATIVE INFLUENCES OF AUSTRALIAN DESIGN

FIG 81

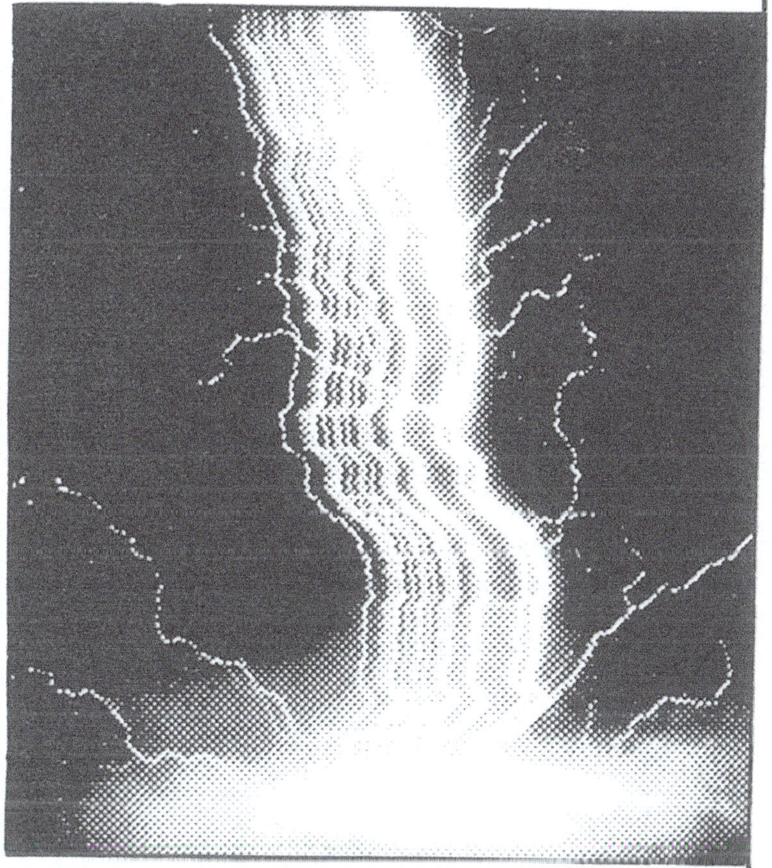

Whilst constructing this Yantram, let this be your meditative Affirmation: RECLAIM YOUR ANCIENT WISDOM

Page 107

Fig 82

Page 108

Construction of the Decagon.

Fig 83

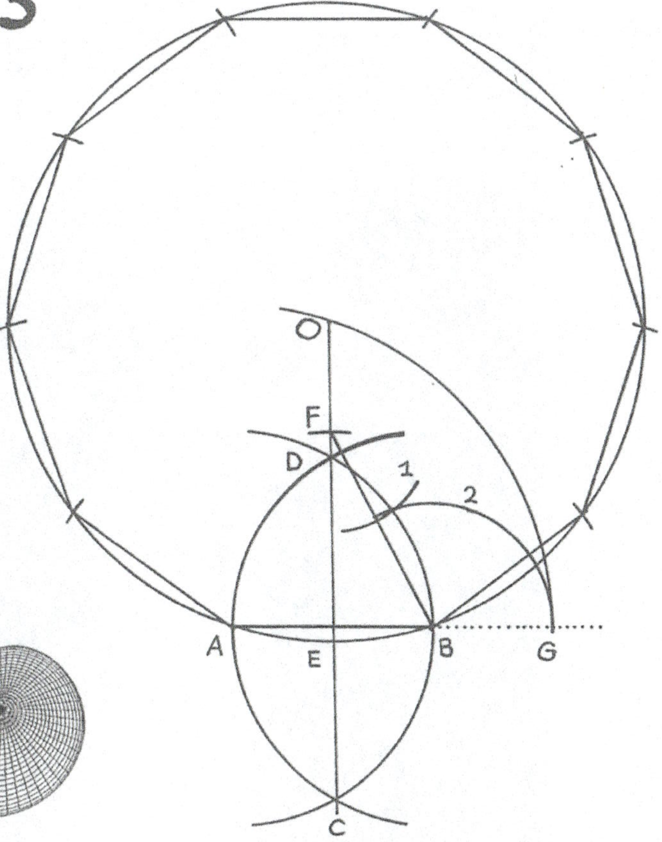

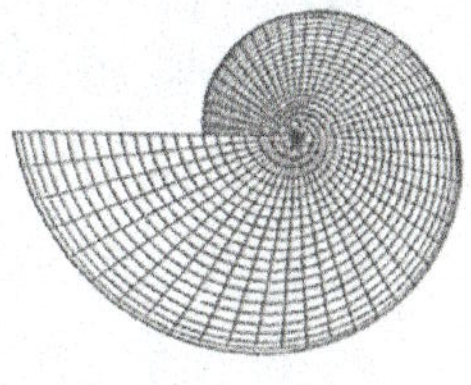

TEN

Fig 84

In the space of Fig 84 you will construct a regular decagon, the same size as shown in Fig 83. Begin with a base line AB which will be the unit edge-length.

Draw the Vesica Piscis (the arcs as shown in Fig 65 for the Enneagon) having radius AB, from points A & B & extend them beyond a ½ Circle. Join the two points of intersection C & D. Extend the centre line CD upwards & from the midway point E, of line AB, measure the unit edge-length to point F & connect F to point B. Take the distance BE as radius & place your compass point on point F & mark off an arc (1). Next place compass point on B & taking the distance along BF from B to the point of intersection made by arc 1, you draw a second tangential arc (2) & extend it to point G on a line extended from AB. Now, taking the distance AG as the radius & with point A as centre, you can draw an arc to meet the vertical extension of line CD at point O which will be the Centre of the Decagon. Take OA or OB as radius & draw the circumscribing circle. Within this circle, given the original line AB you can easily cut off ten equal divisions. Ensure that the lead of your compass is very sharp. & the result will be precise.

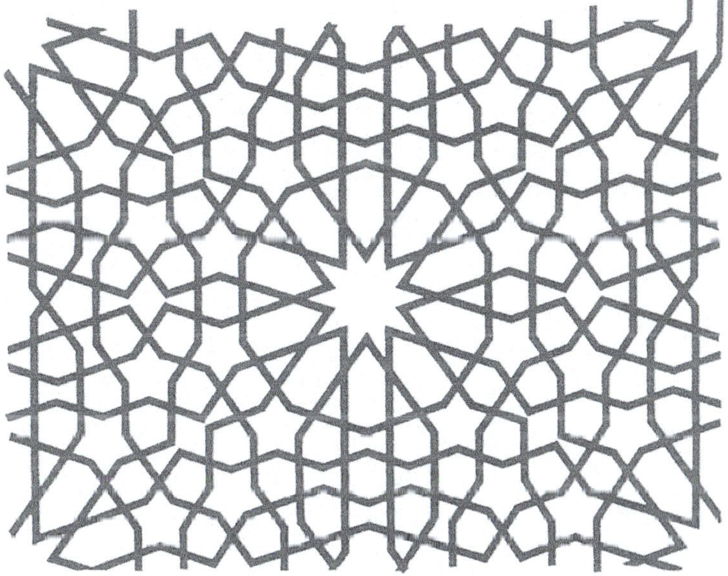

Fig 85

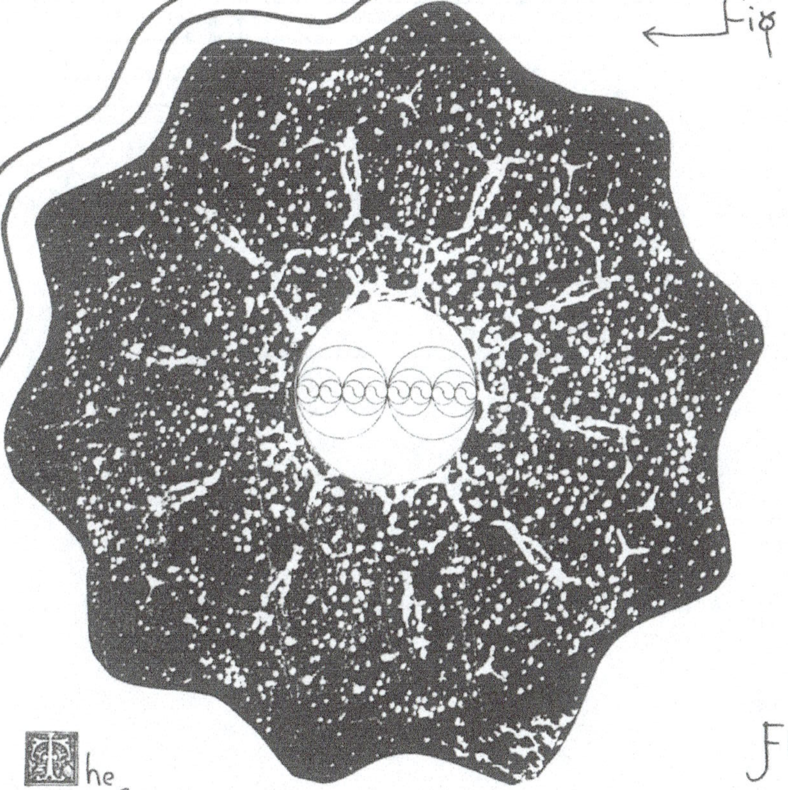

Fig 86

The above fig 85, showing the outline of two concentric pentagrams, is a detailed computer generated model of DNA seen from above. The Two Pentagons form a Decagon. From the same source in Dan Winter's works on 'resonance geometry' is the TEN SPIRALS OF THE GOLDEN MEAN (Fig 86) "A perfect nest of pentagrams. This is the trap which tempts light out of the fast lane, into biology. Light is fooled, it doesn't know whether it is entering a simple net on a planar surface, or an infinite tube. First it finds its way into the nest, there to discover the perfect place to raise the heterodyne children of light without interference".

Page 110

Page 111

THE HARMONICS OF A DECAGON

Fig 87

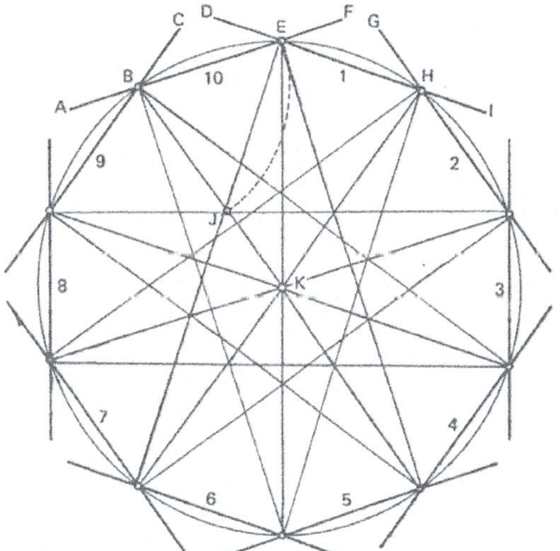

(By Bruce Cathie)

The Decagon indicates a direct relationship between light, the circle, the harmonic power of 10, and the Golden Section ratio of 1.618 (which is the mathematics of biology, Nature, DNA, pyramids, planetary orbits, atomic structure etc).

The subtended angles A-B-C, D-E-F, G-H-I etc each are equal to 144, giving a direct relationship with the harmonic of the Speed of Light. Each inner angle K-B-E, K-E-B, K-E-H, K-H-E etc is equal to 72°, or half the speed of light harmonic.

If the side of a regular decagon inscribed within a circle is laid off along the radius, the radius will be divided into golden section. This is shown in Fig 87 by laying off the side B-E along the radius B-K. The radius is then divided into the ratio of 1 : 1.618, also known as "PHI" whose symbol is "ϕ".

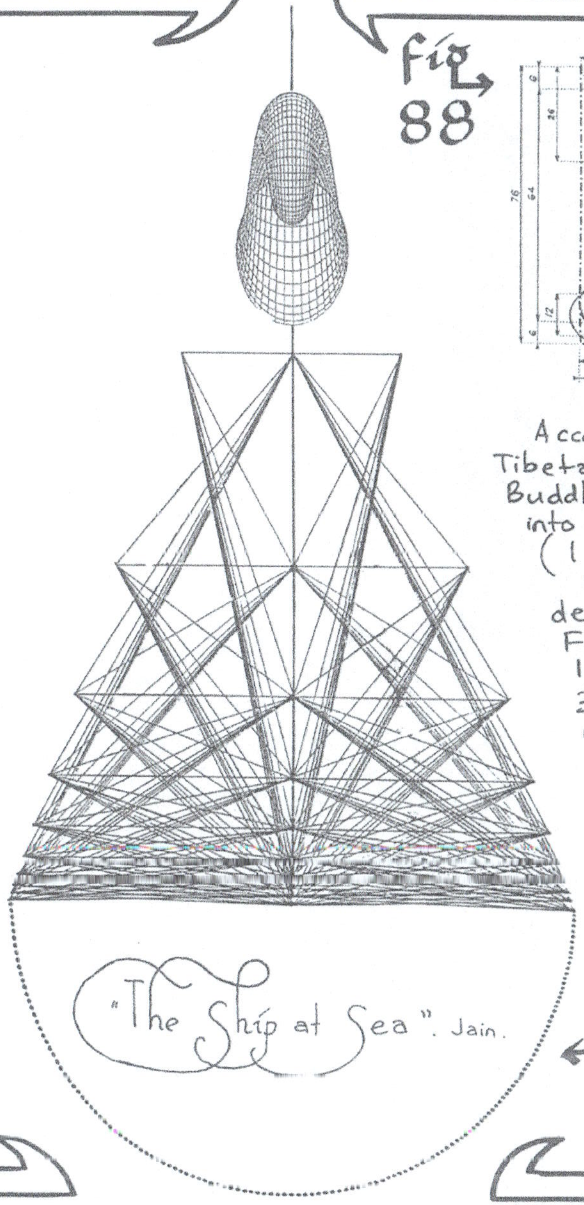

Fig 88

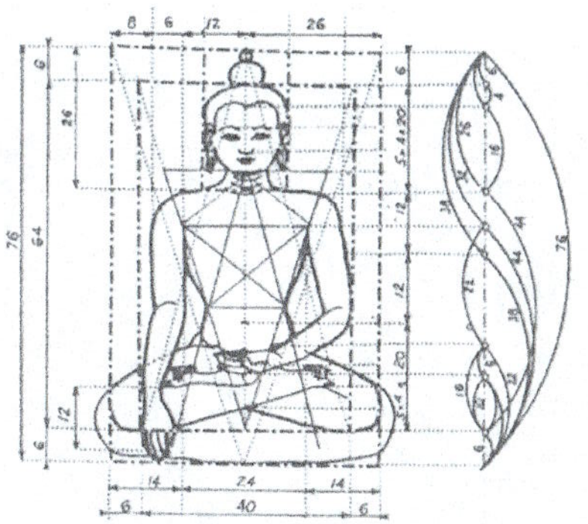

According to ancient Tibetan sources, The Buddha must be drawn into a Golden Rectangle (1 : 1.618) Fig 88!

This ratio can be determined from the Fibonacci series : 1, 1, 2, 3, 5, 8, 13, 21, 34, 55, 89, 144 etc where $\frac{144}{89} \doteqdot \phi$

Fig 89 shows the vertical mask divided into this sequence within a Golden TriΔngle.

Fig 89

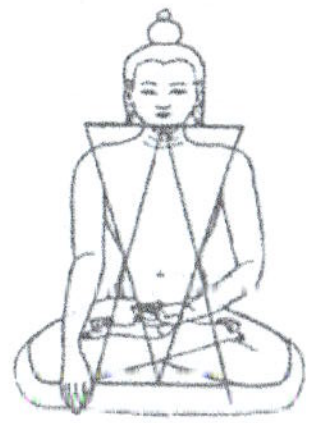

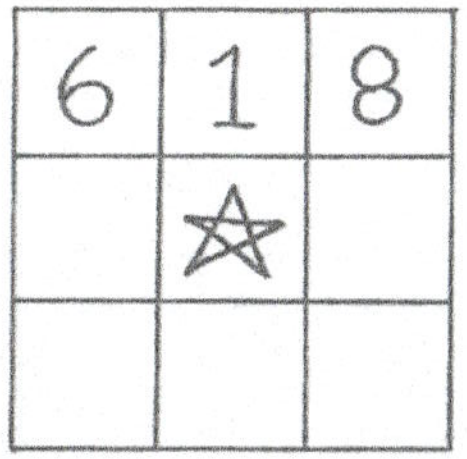

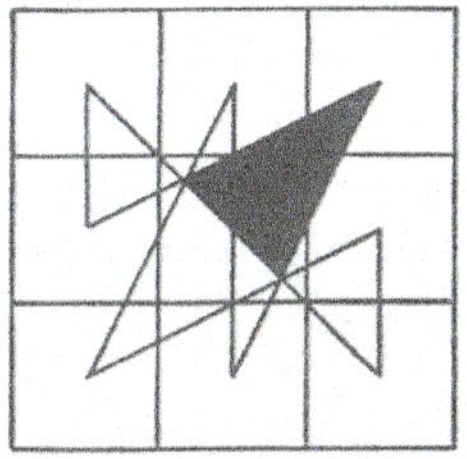

fig 90a ↑

fig 90b ↑

DIVINE PROPORTION

"PHI" Φ

A Study in Mathematical Beauty

fig 90f

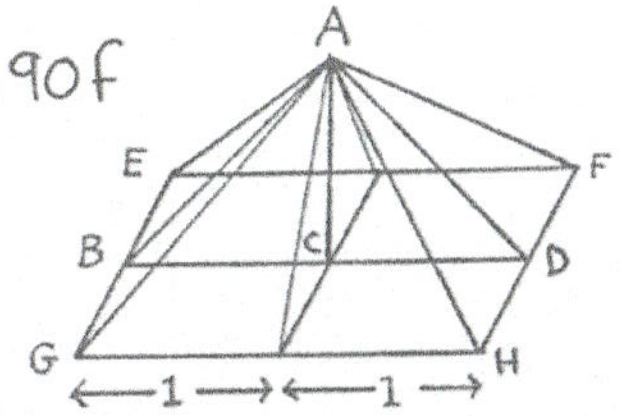

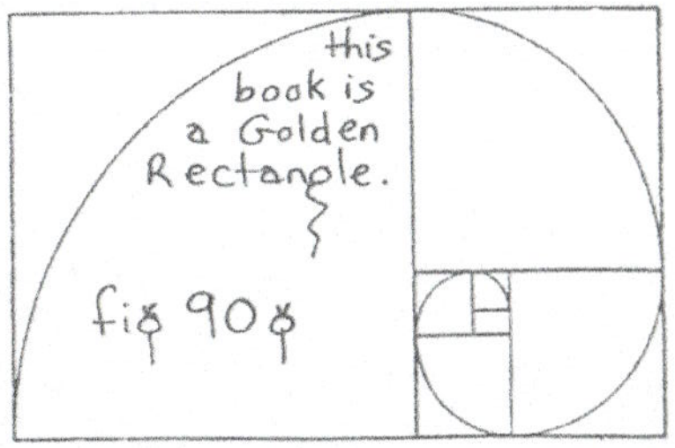

fig 90φ

In my next book I will present a full & detailed description of the Golden Section, but for now, just memorize & chant : "Point Six One Eight" i.e : ·618 : 1 or 1 : 1·618. Lock this into your memory bank & it will trigger more understanding. To stimulate this process I offer you on this page some potent expressions of ·618. Fig 90a is the Majik Square of 3 that has as its top row the expression of ·618. The centre cell is The Cosmic 5 pt star full of Φ proportions. Fig 90b is the yantra of Fig 90a highlighting its Golden Triangle. Fig 90c shows the stars within the stars diminish at the rate of ·618. Fig 90d is my drawing of the Star of Solomon using ·618 Golden TriΔngles. Fig 90e is the Star of Solomon showing AB : BC = Φ. Sacred Art. Fig 90f shows how to fine-tune your pyramid : CD ÷ AD = Φ. Fig 90φ is the Golden Rectangle generating the Logarithmic Spiral or Nautilus Shell curve. Fig 90h is a radiograph of the shell of the chambered nautilus & finally Fig 90i is the Golden TriΔngle's Logarithmic Spiral.

fig 90c →

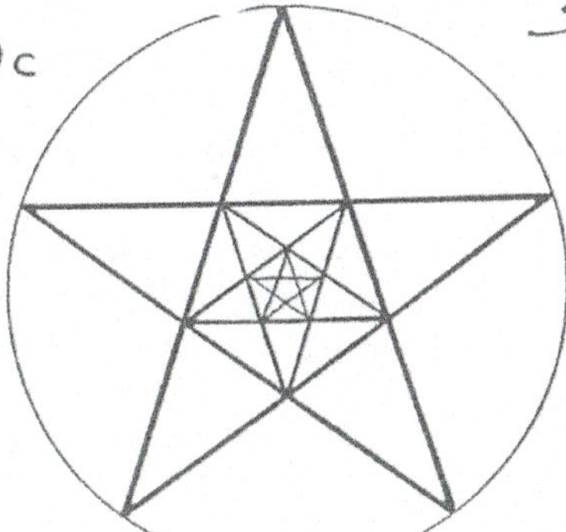

fig 90d

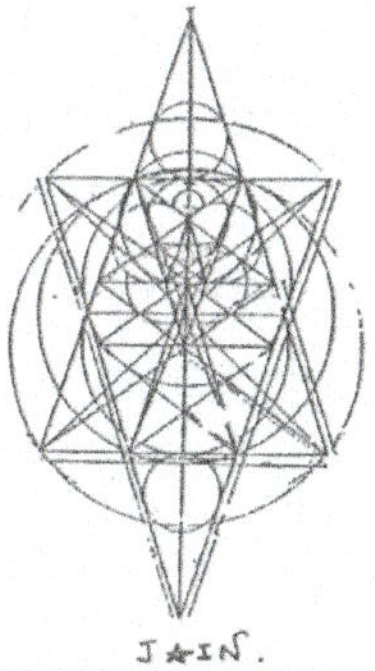

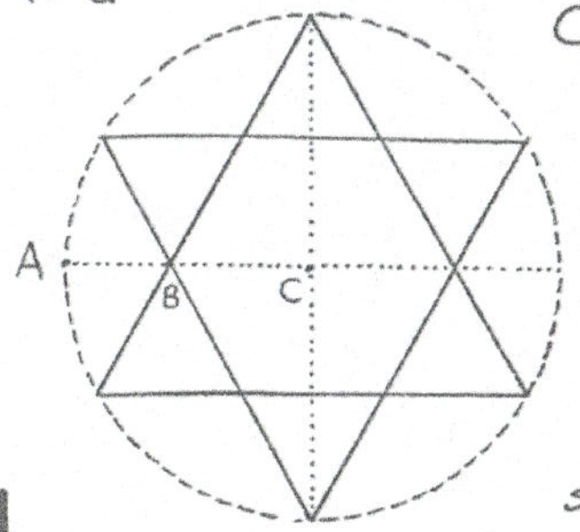

fig 90e ↑

Fig 90h

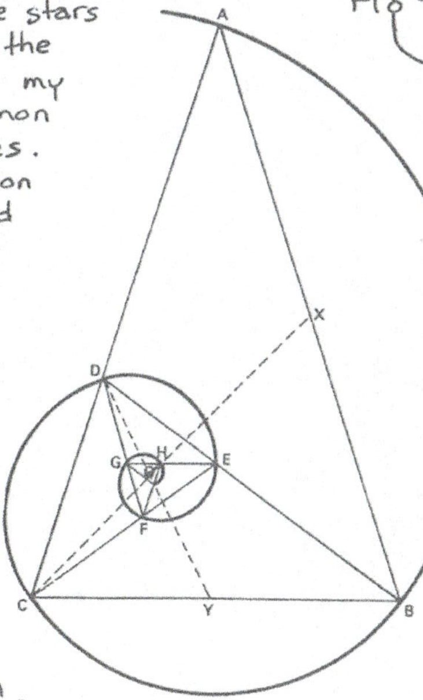

FIG 90i ↑

Page 112

Page 113

A BOOK BLESSING

...(The UNIVERSAL CHURCH of my BEING is located around the wheel of my Heart. This personal mudra of mine (see doodle of the Hands in Prayer) is positioned such that the two joined middle fingers point towards the heart, no matter what position I am in. The two joined forefingers point towards the Highest Reach of ASMI (sanskrit word for "I AM"). This mudra (poetical power & healing positioning of the hands) also appears like a Castle for the Lord God-Goddess of our Beings)... Blessed Be, the God-dess-essence of The-All-That-Is, the Eternal Flame-Spirit withIN me, that has called me to The Light (that which carries Data) & the Love (the Ray of Creation). I illuminate the Supreme Intelligence & the Absolute Life of all existing Harmony/Magick Squares & this I show to the Children of the Stars. This book of divine mathematics is part of the Living Library that I Am & You Are & it is to teach that planet Earth is one of the 12 Inter-Galactic Living DNA Libraries that can only be accessed thru Love of All Life. The Collective Consciousness of these Magick Squares encoded in this Book, is my Altar; the Harmonics of Nature proclaim them, all Life Forms have memory of them, tho' they have been lost & uncared for, I now give rebirth to them. Abraham knew them, Hermes divined them, Pythagoras calculated them, Indian Brahmins examined them, Arabian Monks expanded them, Benjamin Franklin, in every dream of his genius, aspired to them & gave us more. Every culture of repute admired them & I am but a simple sage, a bricklayer by trade, who cares for the Forests & the children of the Earth, wishing the latter a brighter education without the devastating effects of subliminal TV Mind Control, satellite hypnosis, germ warfare of the secret governments, E.T. implants & ego-centered master-geneticists... May the knowledge of this married form of mathematics & Art help us all to over-ride the tyrannies & to be open-minded enuf to realize that even The Tyranny is part of the whole Picture (The Chaos) for it will help wake us all up & start thinking for our-Selves... So as you read thru this book of Harmony Squares, Now, you are coming into the full Light of your own true harmonics, beyond all the deceptions & lies, you are tapping into your HIGHER THOUGHT GEOMETRIES. As you read & draw & understand the wavelengths encoded in these Harmonic Squares, you allow your Self to be genetically changed & accessed to Love. Follow no Gods, but your own SOVEREIGNTY. 卍♡卐☖ Jain. (title of doodle: "Anahata Guru" which means: "MY HEART IS MY TEACHER.").

...The function of this book is to re-introduce Magic Square Artforms as a process of ETHERIALIZATION to subtly launch a revolution in higher consciousness.

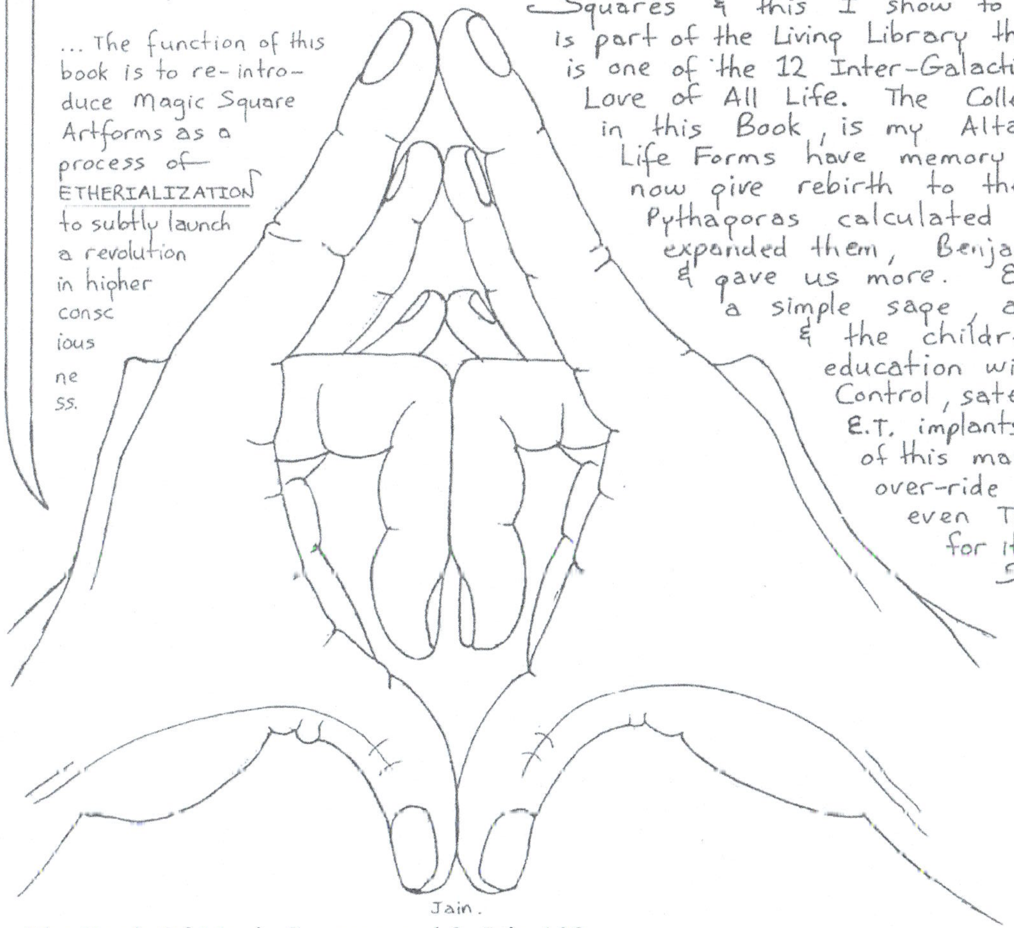

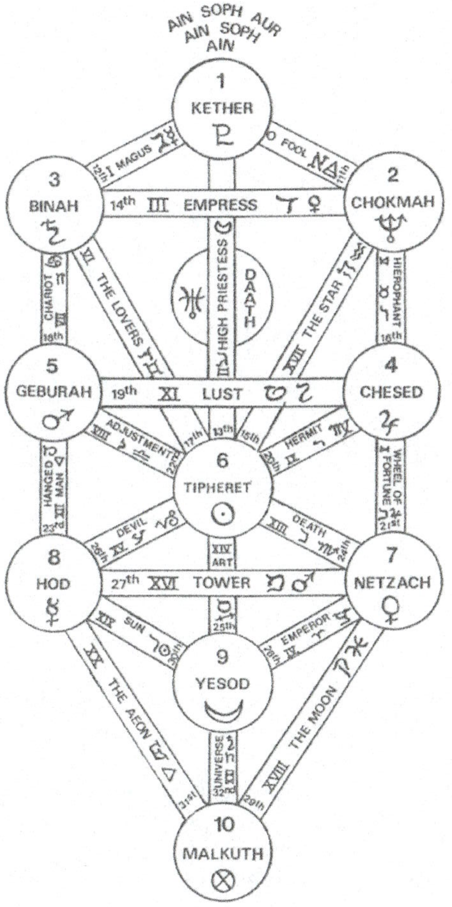

Fig 91

Fig 91 shows the Qabalistic Tree of Life which shows 10 Spheres or planes of consciousness, each one relating to a different planetary vibration. It is like an internal pathway that leads from Earth (Malkuth, Sphere 10) to Heaven (Kether, Sphere 1). Each sphere or Sephirot relates to a specific Majik Square.

Fig 92 is a 10 starred drawing which is the inspiration for your next assignment :

In the space of Fig 93 you will connect each of the 10 dots to every other dot to form a decagonal yantra. Do this in pencil, then proceed to change the geometry by adding or subtracting lines, curves, use of compass, inks, color, writings, runes, etc. Fig 93 is a space for you to be outrageous, unconventional, daring, taking risks, mindless.

mindless

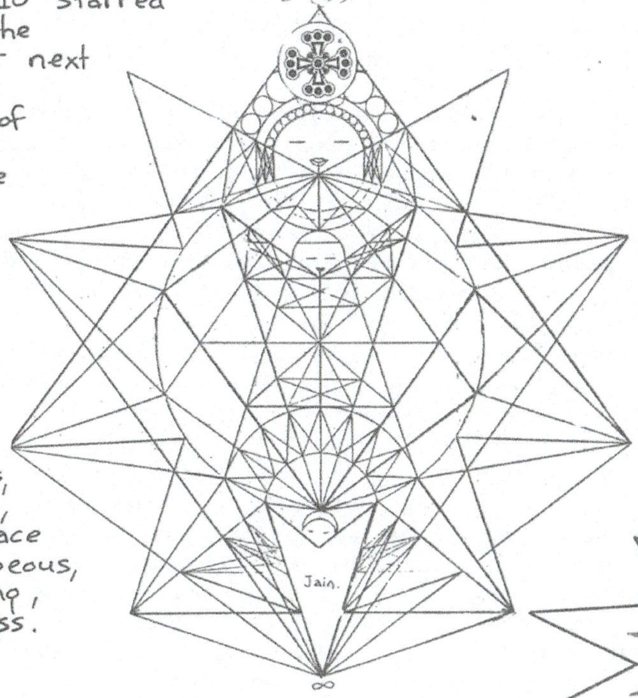

Fig 92

Anglican Cathedral with its beautiful circular & decagonal rose window. →

Convex (◊) or Concave (⧗)?

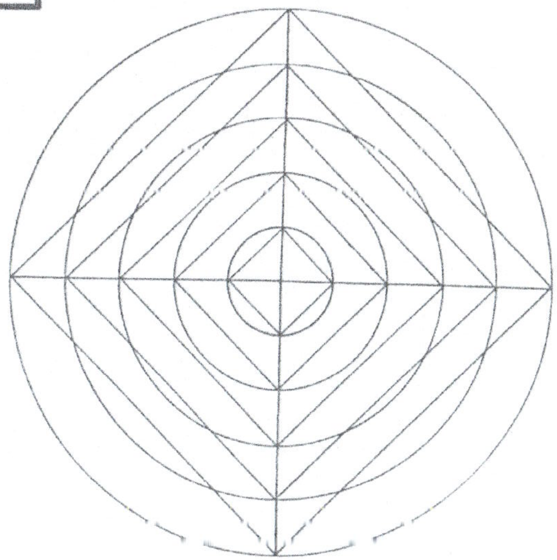

Fig 94

Fig 93

Page 115

Colour this geometrical pattern in black and white or red and white to give the optical illusion that the sides of the outer square are CURVED.

The Ten Puzzle

2	1	3	4
3	4	2	1
4	3	1	2
1	2	4	3

Fig 95

To present you another puzzle, cover up the Nos. of Fig 95 so they are out of view & arrange the Nos. from 1 to 4 in a 4 x 4 square, without repetition of any No. in any column such that :

- The 4 Columns have a sum of 10
- " " Rows " " " " 10
- " " Centre Nos. " " " 10
- " " Quadrants " " " 10
- " 2 Diagonals " " " 10
- " 2 Opposing Pairs " " 10

What is of interest here is to create a design by isolating certain Nos. Here is the position of the 4 ONES. By placing a dot in the centre of the squares we create a Grid (Fig 96) where we can link each of the 4 dots that represent "One". to every other dot.

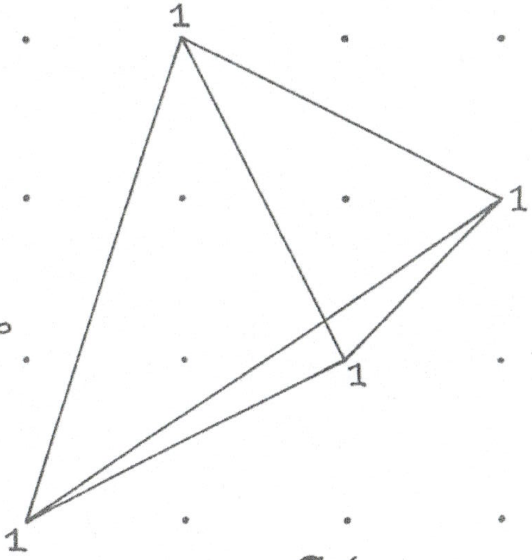

Fig 96

fig 97a

fig 97b

fig 97c

Mindfully connect all the points of "2" in the space of Fig 97a. Refer to the tetrahedral diagram Fig 96 that has emanated from The Ten Puzzle. Similarly connect all the points of 3 in Fig 97b & all the points of 4 in Fig 97c.

Then, in the space of Fig 98, you will combine or superimpose these 4 tetrahedrons (Figs 96, 97a+b+c). When completed, turn this book anticlockwise 90° & repeat the same pattern of the 4 Tetrahedrons upon the same graphic. To do this superimposition, write out Fig 95 & have it infront of you. This Yantra can be referred to as "The Ten Puzzle at 0°+90°"

Fig 98

Page 118

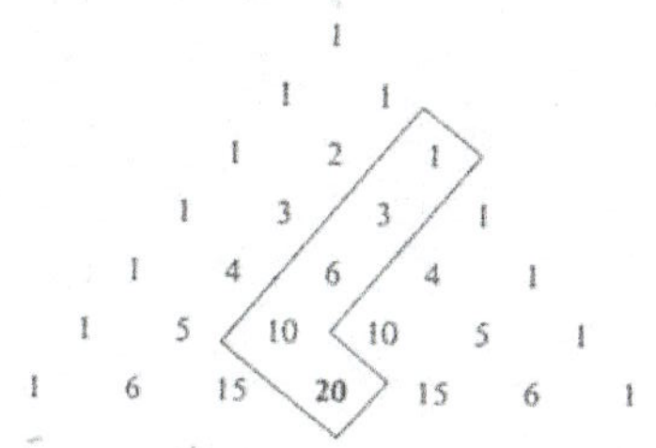

Fig 99 shows that each No. of Pascal's Triangle is formed by the Sum of the two numbers immediately above it.

You can discover for yourself many visual patterns within this triangular array.

The Sum of the numbers of any diagonals in Fig 99 is found immediately below it e.g: 1 + 3 + 6 + 10 = 20 as shown above.

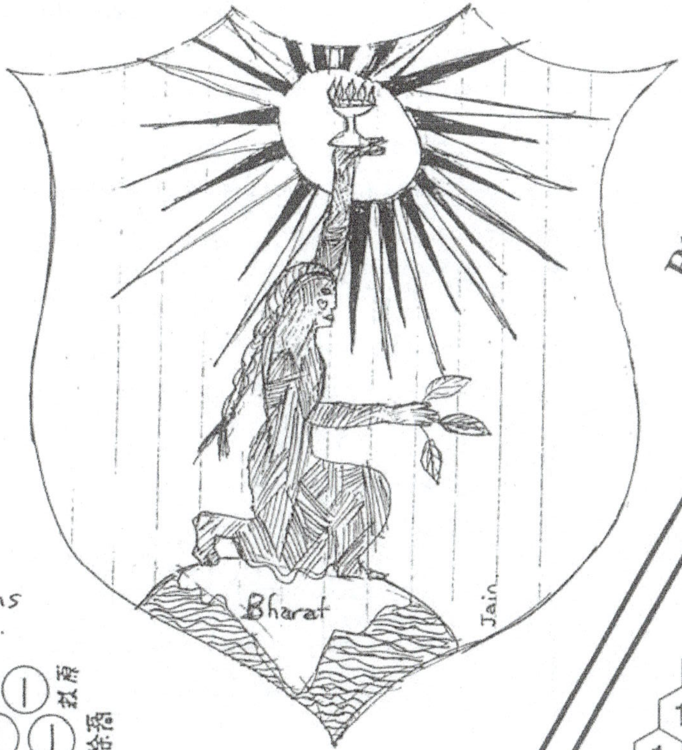

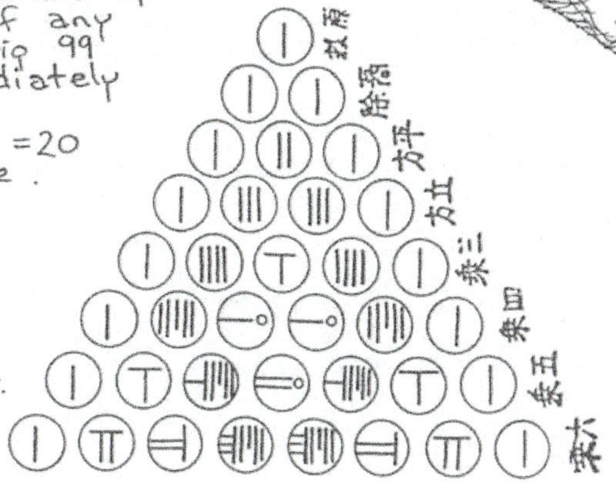

The triangle first appeared in print on the title page of an arithmetic book by Petrus Apianus at Ingolstadt in 1527.

Pascal's Triangle, similar to the one shown above, was shown on a Japanese mural in the 1780s. The Chinese were aware of the Triangle in 1303.

Fig 99

Page 119

VISUAL PATTERNS IN PASCAL'S TRIANGLE

Multiples of 10

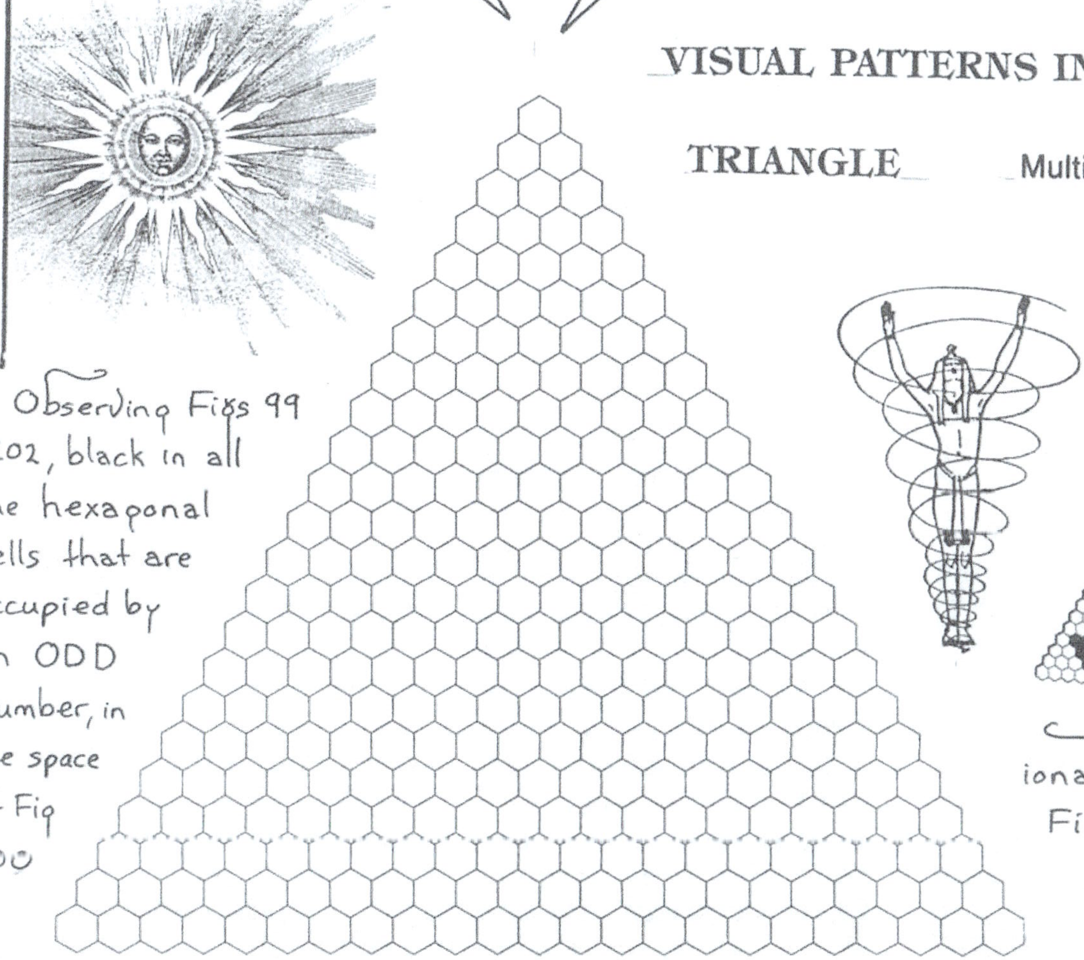

Observing Figs 99 & 102, black in all the hexagonal cells that are occupied by an ODD Number, in the space of Fig 100

Fig 100

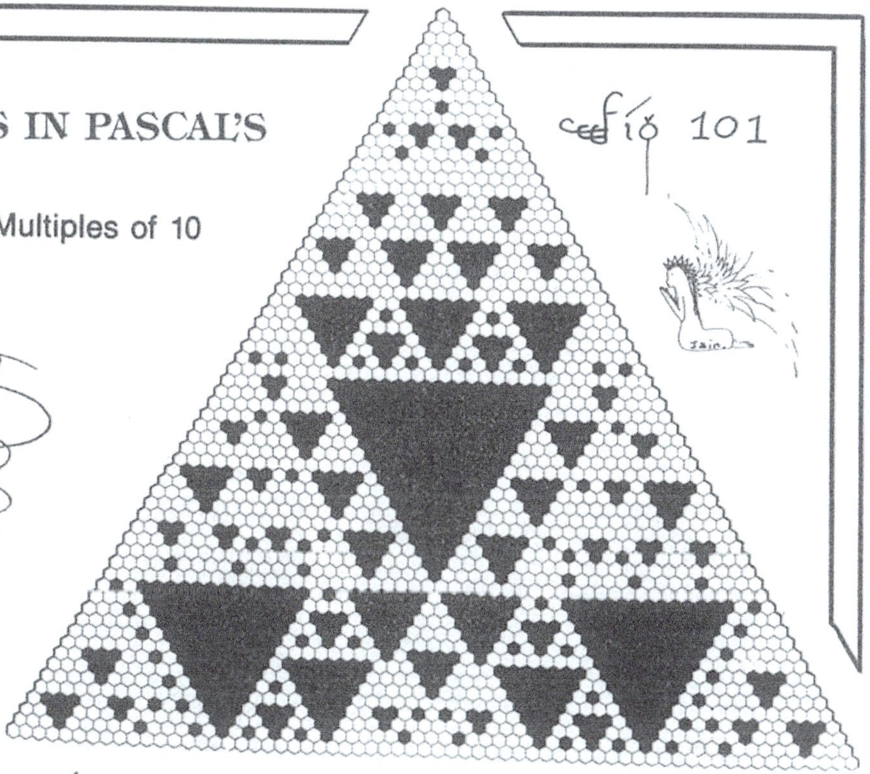

Fig 101

An example of this creative & explorational internal pattern search is shown in Fig 101 where all the multiples of 10 are blacked in. Let the next exercise be categorized as Fig 102, & that is to fotocopy Fig 100 9 times, & on 9 separate grids, black-in all the Multiples of 2, 3, 4, 5, 6, 7, 8, 9 & 10. See 'Solutions' on P21.

Page 120

B R E A T H E

PEACE OF MIND

Fig 103

Seat-of-the-Soul Chakra. The place where the soul is anchored to the etheric (memory) and physical bodies. This is the chakra of freedom, the violet fire of freedom, which is the seventh ray of Saint Germain (master of the imminent seventh, or Aquarian, age). Through the science of the spoken Word, you can direct the accelerating violet flame of the Holy Spirit into the accumulation of karma recorded in the subconscious 'electronic belt'. And you can feel how the flame renews thought and feeling, liberating the soul to become all that God made it to be.

The six petals of the soul chakra represent the six-pointed star of victory. They govern the flow of light and the karmic patterns in the genes and chromosomes and in the sperm and the egg of man and woman.

ECP

Solar-Plexus Chakra. The "place of the sun." The solar plexus is the center of feeling where we employ the energy of emotion as God's *energy in motion* to realize the peace of God's consciousness, the peace of the ascended Jesus Christ. When your solar plexus is calm, you have the power of peace. And, using the full potential of the desire body, you have that full momentum of the "Pacific Ocean," of water molded in a matrix of love.

The flame of purple and gold is God's desiring to be God within you. And the power of God's desire comes forth as you raise the energy from the solar plexus to the level of the throat and release the fiat, as Jesus did, "Peace, be still!" Through the spoken Word, a wave of light goes over the whole planet, causing atoms and electrons to come into alignment with the flame of the Prince of Peace. The solar-plexus chakra has ten petals.

ECP

Throat Chakra. The power center in man and woman. The sixteen petals, or frequencies, correspond to the thought form of the pyramid (four petals on each side), which represents the power of precipitation—the Word becoming flesh, Spirit becoming tangible in Matter.

The throat chakra is the center of the blue flame and the inner blueprint of the will of God. It is the key to the shortening of the days for the elect that Jesus Christ said would occur (Mark 13:20). The shortening of the 'days' or cycles of karma occurs through the correct use of the spoken Word. When we speak the name of God "I AM" and follow it with affirmations of light, we begin the transmutation process. Anything and everything that proceeds from the throat chakra coalesces in form, for good or for ill, by the action of the power of the spoken Word. Its scientific use is the Truth that shall make you free when you apply it diligently every day.

ECP

Third-Eye Chakra. We return to the consciousness of absolute Good through the third-eye chakra, which has ninety-six petals. The third eye, vibrating in the emerald green of the science of truth, gives us the immaculate picture of individuals, of civilizations, of the divine pattern.

Jesus said: "The light of the body is the eye." (Matt. 6:22.) Through the all-seeing eye of God, the inner eye of the soul, you tune in to what should be in reality, instead of what may be occurring in the actualities of the present. You can always tell whether or not you are looking through the third eye or through the two eyes: the third eye always gives you the immaculate concept of the blueprint of Life as well as the discrimination to know Good and evil. The two eyes give you a relative and often unclear perception and perspective on life. They do not penetrate beyond the physical plane unless in an exalted vision the faculty of sight be accelerated by miraculous intercession.

ECP

of fig 103: In Sanskrit, this Solar Plexus chakra (lit: "Wheel of Energy") is called MANIPŪRA (lit: Jewel-Filled), relating to the element of Fire. It is at this stage of Evolution-Involution that Man/Woman needs to get into hir emotions; especially that men learn to Cry, only then can we be True, Whole, Creative.

From Historia Deorum Fatidicorum.

PYTHAGORAS, THE FIRST PHILOSOPHER.

During his youth, Pythagoras was a disciple of Pherecydes and Hermodamas, and while in his teens became renowned for the clarity of his philosophic concepts. In height he exceeded six feet; his body was as perfectly formed as that of Apollo. Pythagoras was the personification of majesty and power, and in his presence all felt humble and afraid. As he grew older, his physical powers increased rather than waned, so that as he approached the century mark he was actually in the prime of life. The influence of this great soul over those about him was such that a word of praise from Pythagoras filled his disciples with ecstasy, while one committed suicide because the Master became momentarily irritated over something he had done. Pythagoras was so impressed by this tragedy that he never again spoke unkindly to or about anyone.

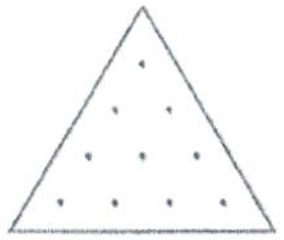

THE TETRACTYS.

Theon of Smyrna declares that the ten dots, or tetractys of Pythagoras, was a symbol of the greatest importance, for to the discerning mind it revealed the mystery of universal nature. The Pythagoreans bound themselves by the following oath: "By Him who gave to our soul the tetractys, which hath the fountain and root of ever-springing nature."

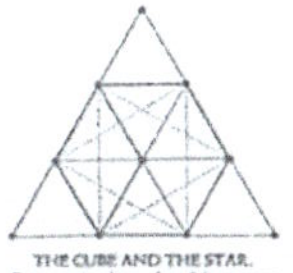

THE CUBE AND THE STAR.

By connecting the ten dots of the tetractys, nine triangles are formed. Six of these are involved in the forming of the cube. The same triangles, when lines are properly drawn between them, also reveal the six-pointed star with a dot in the center. Only seven dots are used in forming the cube and the star. Qabbalistically, the three unused corner dots represent the threefold, invisible causal universe, while the seven dots involved in the cube and the star are the Elohim—the Spirits of the seven creative periods. The Sabbath, or seventh day, is the central dot.

PYTHAGORAS

Born in Sidon, Lebanon, Pythagoras travelled extensively for 22 yrs before he started his community in Crotona, Italy. né circa 590 BC.

Like the Tibetans, he believed that each number had a Soul, a living unique vibration, & thus Numbers were Holy. This sacredness was expressed wholly in the "TETRACTYS" shown above as fig 104. 'Tetra' means 4, thus the sum of 1+2+3+4 = 10 & this triangular arrangement of "10"ness revealed to the Community of Sacred Geometers that Number is a clear window into the Mystery of the Universe. The Pythagoreans bound themselves to the following oath: "By Him-Her who gave to our soul the Tetractys, which hath the fountain of ever-springing nature".

fig 104

Pythagoras, the Philosopher of Samos (Greece) is the father of musical scales. He took the stretched string & with successive midpoint pluckings developed a whole cosmological musical scale that we still use today.

"5" was the union of the first even (2) + odd (3) numbers & represents Marriage. "10" represents all the principles of divinity, evolving & re-uniting in a new unity (As 10 = 1+0 = 1). Eventually, all Pythagorean ascetics were brutally murdered.

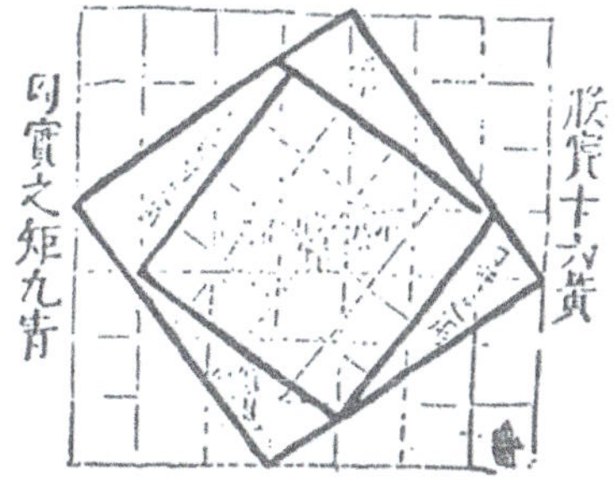

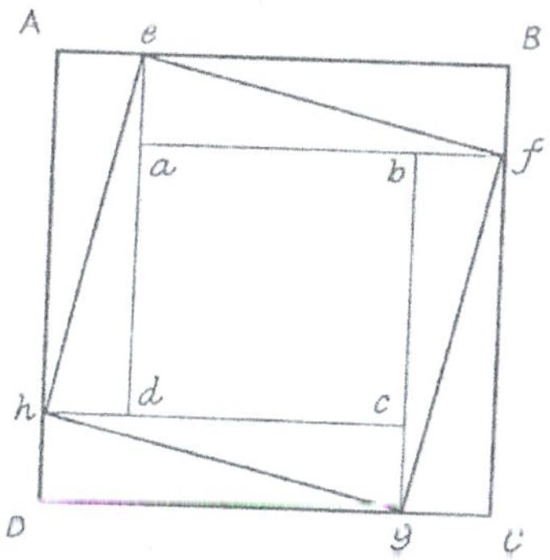

Fig. 18. Chinese Proof of Pythagoras' Theorem

The Book of *Chou Pei Suan King*, probably written about AD 40, is attributed by oral tradition to a source before the Greek geometer taught what we call the *Theorem of Pythagoras*, i.e. that the square on the longest side of a right-angled triangle is equivalent to the sum of the squares on the other two. This very early example of block printing from an ancient edition of the *Chou Pei*, as given in Smith's *History of Mathematics*, demonstrates the truth of the theorem.

PYTHAGORAS OF CROTONA.

Though Pythagoras excelled Plato in the profundity of his philosophic deductions, the transcendental element in his doctrines has provoked the ridicule of a materialistic modern science. All too lightly the world has passed over the achievements of the first "philosopher" to whom it is indebted [illegible]. The twentieth century student of Greek philosophy has been taught to associate the name of Pythagoras with such puerile issues as his golden shin bone and his ban upon the eating of beans! It has also been held against Pythagoras that he delivered his discourses from behind a curtain, preached in allegories and enigmas, and revealed his scientific knowledge only to such initiated disciples as had spent many years in self-discipline. His scholastic critics, however, have failed to consider the manner in which Pythagoras obtained his immense wealth of abstract learning. The Mysteries of Greece, Egypt, Persia, and India without exception bound their initiates to inviolable secrecy. Having accepted the obligations of these societies, Pythagoras had no honorable course other than to abide by their regulations. Iamblichus lists 218 men and 17 women among the most famous of the Pythagorean philosophers. It is thus evident that Pythagoras revealed his secrets to a considerable number of persons—probably all who he felt could understand and be benefited by his knowledge. The Pythagorean doctrine of mathematical philosophy may yet be accepted as the one system of thought able to cope with the riddle of existence.

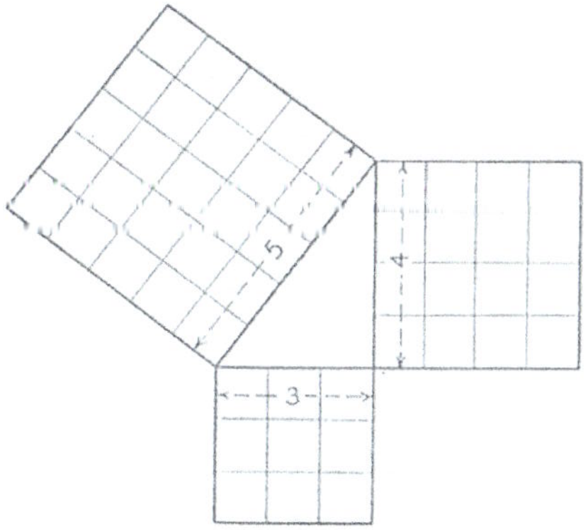

Fig. 17. The Right-angled Triangle of the Temple Builders

Long side:	5 feet	($5^2 = 25$ *square feet*)
Short sides:	4 feet	($4^2 = 16$ *square feet*)
	3 feet	($3^2 = 9$ *square feet*)

$25 = 16 + 9$ *or* $5^2 = 4^2 + 3^2$

Page 121

Page 122

The Squaring of the Circle

or "Creation of the Pentagon"

"The Mystic Quadrature of the Circle"

The task is to make the perimeter of the Square (Manifestation) to be equal with the Circumference of the Circle (Spirit). Symbolically it means Heaven (Circle) on Earth (Square). The Final Geometry is also the secret to the Creation of the Pyramid proportions as shown in Figs 108 & 109.

In the World of Islam, the concept of Squaring the Circle is enacted today ritually by the circumambulation of the Kaaba (a large black Cube that has Mohammed buried within) at Mecca the Holy Cube.

Because the circle's formula is based on π (= 3·142 etc ∞) which is irrational ie not based on whole numbers, the squaring of the circle can only be approximate. See next page for the step-by-step Construction.

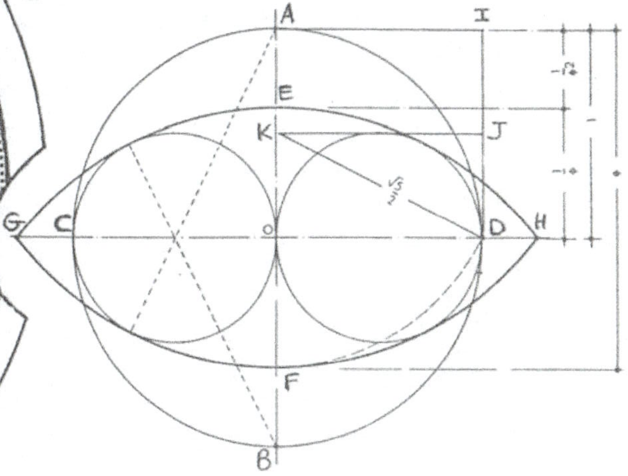

Fig 105

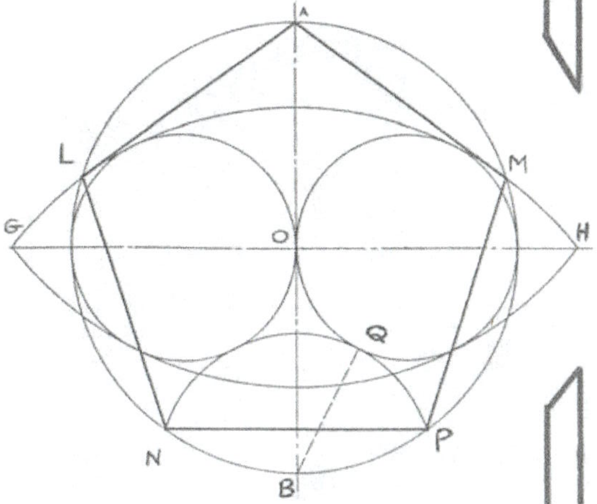

Fig 106

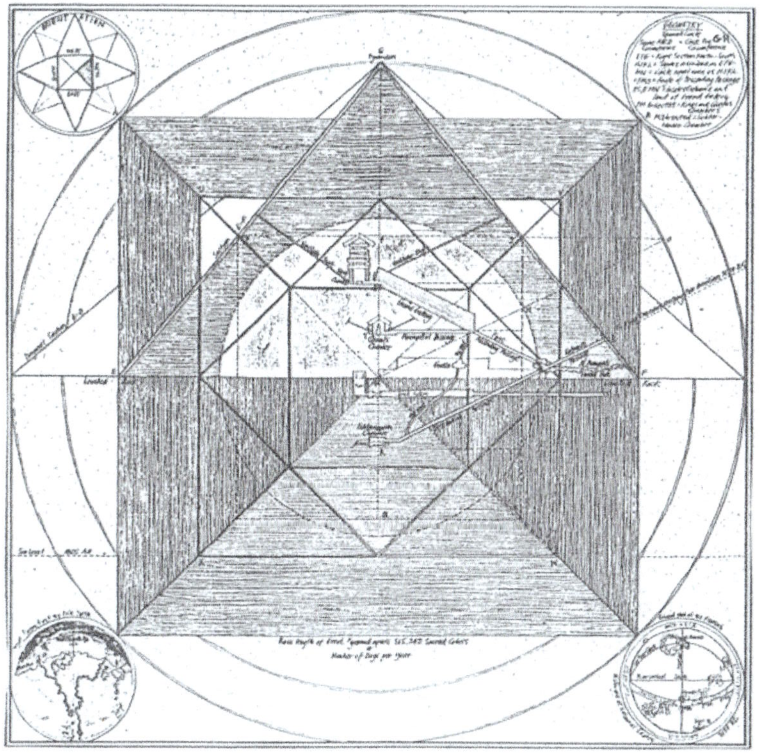

Fig 109

To begin Squaring the Circle we first create the Pentagon (Fig 106).

Draw a circle. O = Centre. Radius OA = 1 Diameters AB & CD are at right angles. Draw 2 circles each having a radius of ½ the original circle, their centres upon diameter CD. Place compass on A & swing an arc GH being tangent to the circumferences of the two smaller circles. Do the same from point B. Construct square AIDO from the original circle's radius OA. [In one simple move, the radius OA = 1 can be made into its Golden Proportion of $\phi = 1.618$ by swinging down the arc of the semi-diagonal DK of the square AIDO, to intersect diameter AB. Point F is the Golden Proportion. Also, the arc GEH divides the radii AO & BO into the Golden Division of $1/\phi = .618$ and $1/\phi^2 = .382$ respectively].

Notice how fig 105 creates the Yin-Yang Symbol. One becomes Two. The 2 circumferences of the two inner circles are equal to that of the larger circle! The area of the two is ½ that of the original! The Homogeneous Unity has become the Polarized Duality. The arcs GEH & GFH are known in ancient times as The VESICA PISCIS, a Godly symbol of Creation.

Regarding Fig 106, from point A draw a line to L which is tangential to the intersection of the Vesica Piscis + inner circle. Then draw AM. So far, A, L & M are points of the Pentagon. Place compass at B, draw arc NQP that touches the 2 inner circles, to create the points N & P. Connect ALNPM to form the Pentagon, the ultimate symbol of life whose symmetry appears only in living organisms + Nature's garden.

Observing Fig 107, enclose the initial circle in a square RSTU. Then place your compass at the centre of the initial circle at O, & with radius extending to the tip of the Vesica Piscis at G & H commence to draw the final circle, whose radius = OH or OG. Now you have created the Mystical Squaring of the Circle where Sq.RSTU = circumference of Circle. Now that the secret is out, herein lies the wisdom of the true Pyramid angles. Let base of Pyramid = CD & extend slope heights C to V & D to V as shown in figs 107, 108 & 109. Commit this information to memory & use it only for sacred purposes in art & building. ॐ

Fig 107

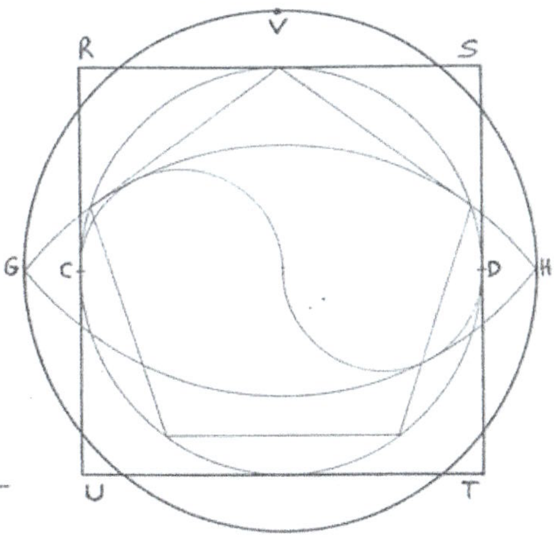

Fig 108

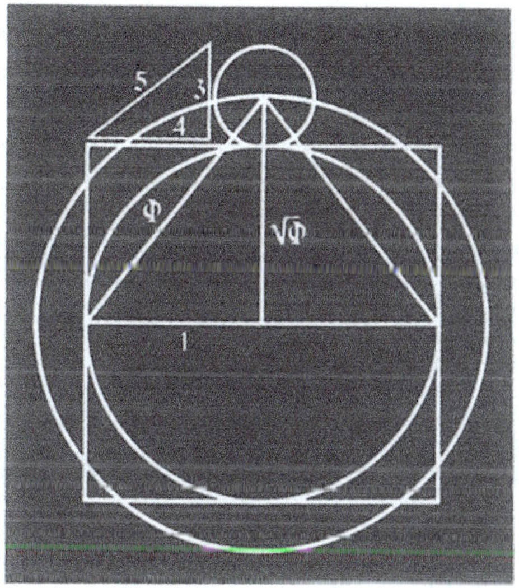

Page 123

Page 197

Chapter Zero: "Solutions to the Harmony Squares".

SPECIAL EDITION

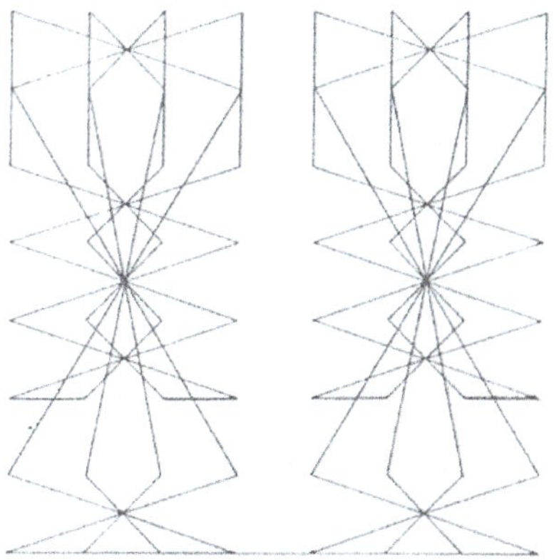

fig 3

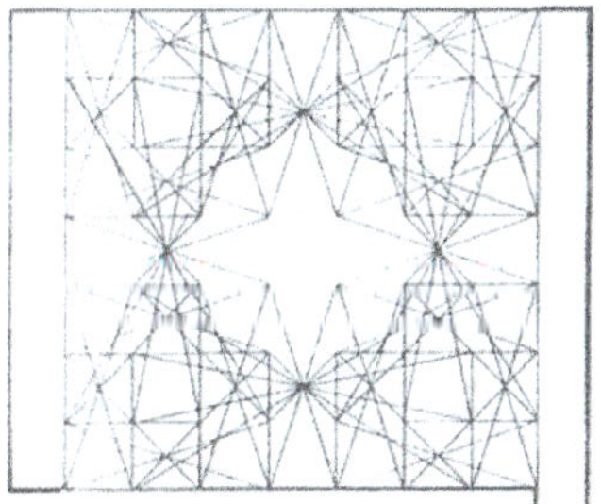

Page 198

Don't take all this
sacred geometry too
seriously, 'cos the other
point of view is to have
your Circumference
no-where & your
Centre
Every
w here.

"Nowhere" =
NOW HERE.

Beauty in Mathematics

"BEAUTY IS A WORD OF GOD"

The Lost Journals of the

HARMONY SQUARE Magus

If this be the last
book I write for a long while
it is because I
have become a
recluse, a monk
of Majik
Squares.

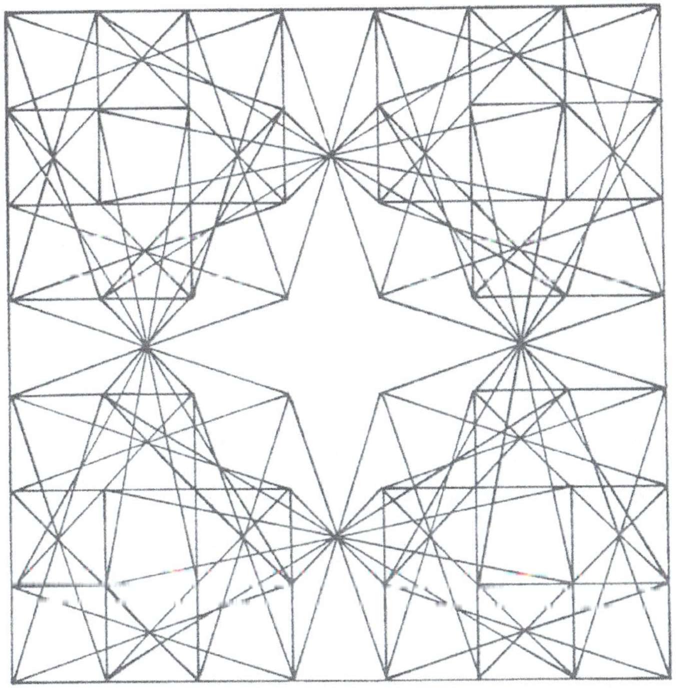

FIG 8

Page 199

4	5	4	5	4	5	4	5
6	3	6	3	6	3	6	3
5	4	5	4	5	4	5	4
3	6	3	6	3	6	3	6
7	2	7	2	7	2	7	2
1	0	1	0	1	0	1	0
2	7	2	7	2	7	2	7
0	1	0	1	0	1	0	1

FIG 11

SOLUTIONS

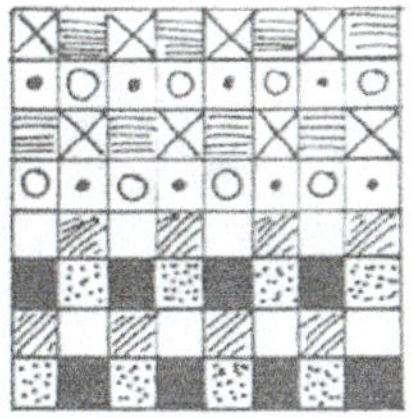

"The Creator asks not to be worshipped in an external image, but to be acknowledged in each human being".

Fig 12

Fig 10

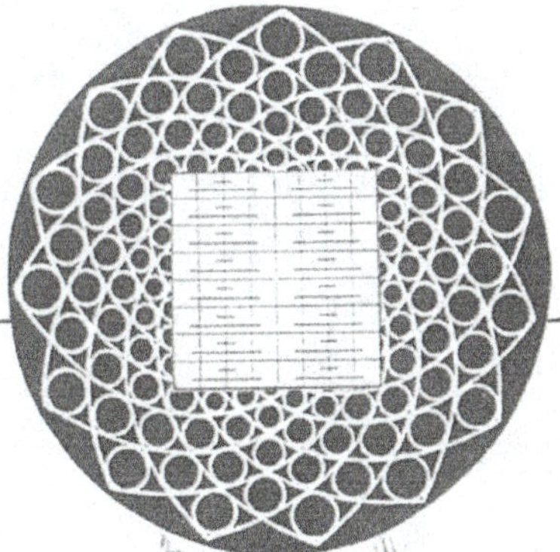

Fig 9

"I AM A KEEPER OF FREQUENCIES". Different M.Sqs have different sums. This Sum or CONSTANT represents a certain Frequency. You will observe, in Fig 13, that M.Sqs. can be decoded into brain-wave patterns. That is why I am giving out these squares or frequencies, not to teach you, but to awaken you, to reMIND you, to inspire you. There is a time when information like this must be guarded & kept secret, but there are times like now at the end of this Kali Yuga (Age), when such knowledge must be omni Available / Accessible. What better way to access these frequencies than a children's / adult's book of self-drawn, self-colored designs. Just by drawing a M.Sq then coloring it in, impregnated with positive Thoughts / Affirmations, is a clear method of contacting that specific frequency that you may need for Healing, Relaxation, Harmony amidst Chaos/Conflict. By Harmony, I mean an orderly & pleasant joining of diversities. Different elements do complement each other by Joining. The Greek origin of the word "HARMOS" is "TO JOIN", thus "Harmony Squares" rather than "Majik Squares". The concept of this book, like Fig 3, is TO JOIN THE DOTS to arrive at an abstract icon of Harmony. Many children's first drawings in pads were series of dots to join, being numbered sequentially, to form animals, insects etc. Same principle. Simplicity.

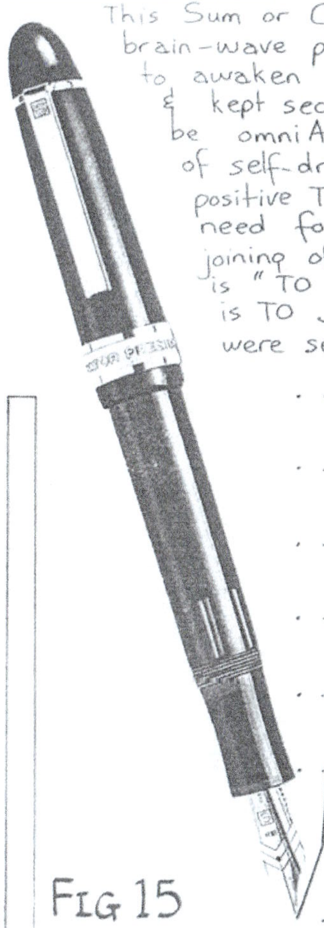

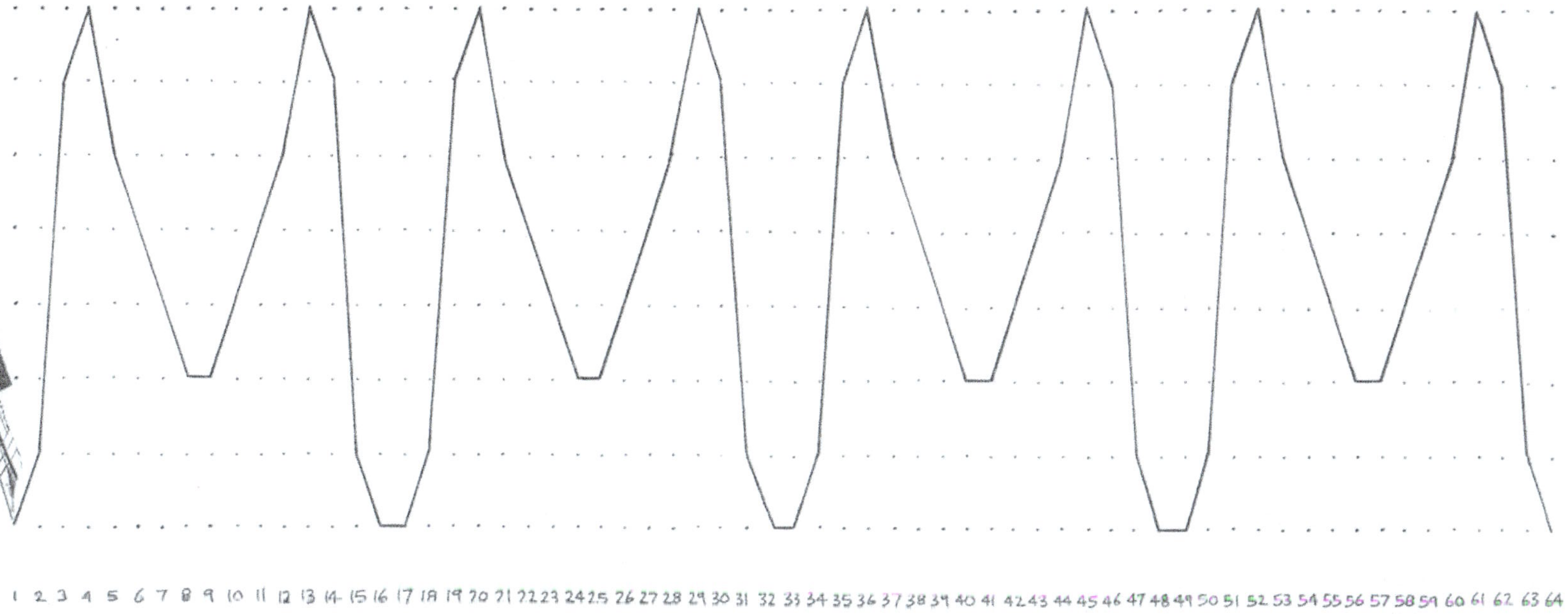

FIG 15

Page 200

Each individual piece of lifes puzzle, no matter how large or small, helps to clarify the over-all picture once that piece is fitted into place.

This is a self-portrait doodle from 1982. I have here in my studio over 1000 doodles & over 1000 very geometrical original Majik Square Graphics. Now, you would think to create 2 separate books: this is what I once thought: "The Book of Doodles" & "The Book of Magic Squares". So, as you can see, the evolution of my drawings & designs were not to separate them but to combine them, as I did in Vol 1 & now in Vol 2. You might ask: What do you do with all this stuff? The answer is that the need to print all this stuff is like a grand catharsis; if I don't wade thru & utilise this hoarded material, I may as well dispose of it or burn it. Thus I am lightened, for this book is my healing, my karma, my devotion, my ticket to travel around the world teaching colorful & exciting new mathematics of Nature & our Ancient Elders; to re-weave the magic golden thread of 1:·618-ology. You might also ask, what's a page like this doing in this Chaptero Zero: "Solutions"? well, I don't know how it happened, there was a technical mistake in the numbering of the pages & I had this blank page. So I stuck this doodle in & began to write these words, realising, this doodle has been rejected for a decade, so I might as well as use it, 'cos I've run out of my best graphics that were blue-tacked on the wall & it is THE SOLUTION: This method of printing & publishing my combined doodles + geometry IS The Solution, at this delicate stage of my present Incarnation. Its like each doodle has its own Consciousness like a carrot in your garden begging you to pluck it & have it for lunch. When I drew this doodle (in the Torres Straits 1982) I was looking in the mirror sketching mySelf. Its cosmic. It means we have to look at ourSelf, to know any Solution, to study ourSelf, know & Love ourSelf... Then you might ask why am I writing at this angle? why am I writing simple-minded? why am I being different? because I'm just being me, I'm not even trying to change your perception 'cos it will change anyway. And its good not to be too pedantic & intellectual & serious, just easy... And what does that symbol "♆" drawn in the book of the doodle, mean? This "♆" is another accepted version of "♆" or Neptune. Astrologically, Neptune is the mystical, spiritual, dreamy, illusionary, nebulous energy within all of us. I am very Neptunian, born at sunrise when ♆ was on the East (+5 planets in the 12th House). ♆ rules the 12th House which relates to the 12th sign of the Zodiakos – Pisces... and a study of astrology led to my recognition of the magic squares which are planetary energies. So you would benefit by studying astrology... whereas, someone like Jimi Hendrix was born with an intensely dominating influence of Uranus (♅). So you could say that Jimi's musik is pure Uranian Energy: revolutionary, new, exciting, berzerk, electrical, breaker down of traditional values & barriers, space age, New Age

The
NOW AGE

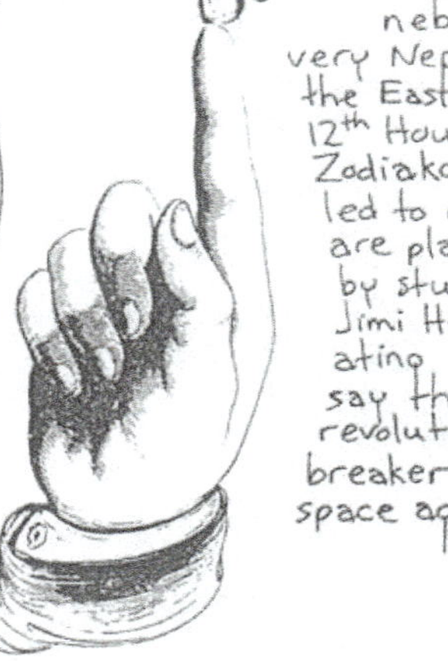

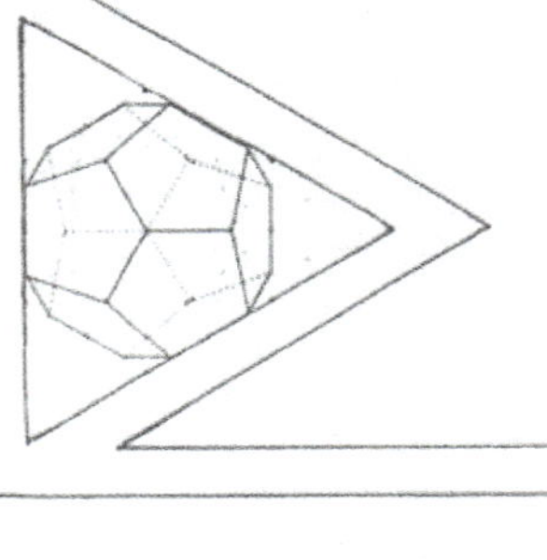

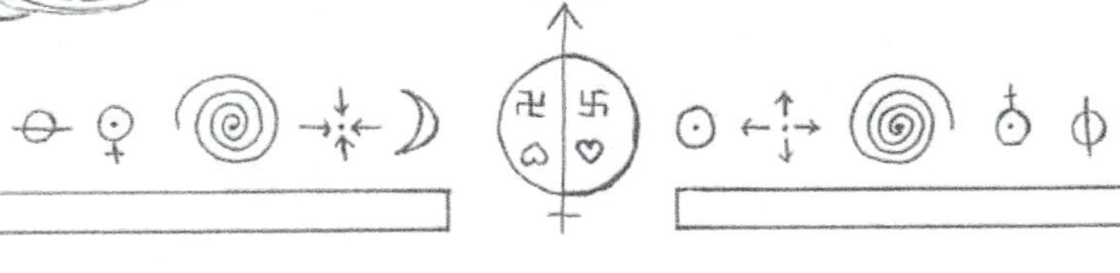

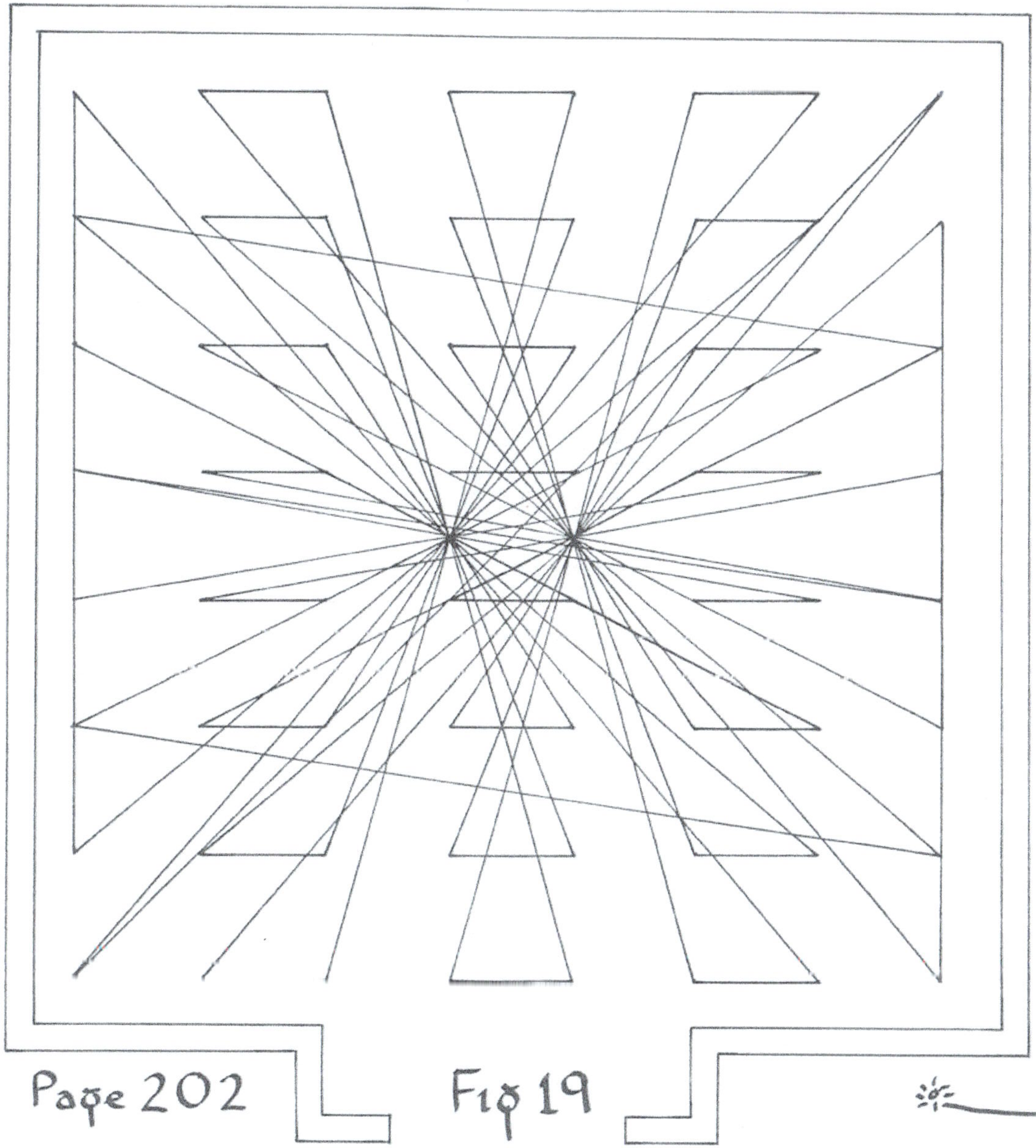

Page 202 Fig 19

Some esoteric books may relate the majik □ of 3 to Saturn, & of 4 to Jupiter etc. I do not know why; and different books attribute other planets 2 other majik squares. I ∴ make no equations. As far as I am concerned, m.sqs do relate to planetary vibrations, but until I am enlightened about IT ALL, I keep chanelling, & just "bliss-in" on the designs of Free Energy.

Paracelsus wrote that each of us has two bodies, one compounded of the elements the other of the Stars.

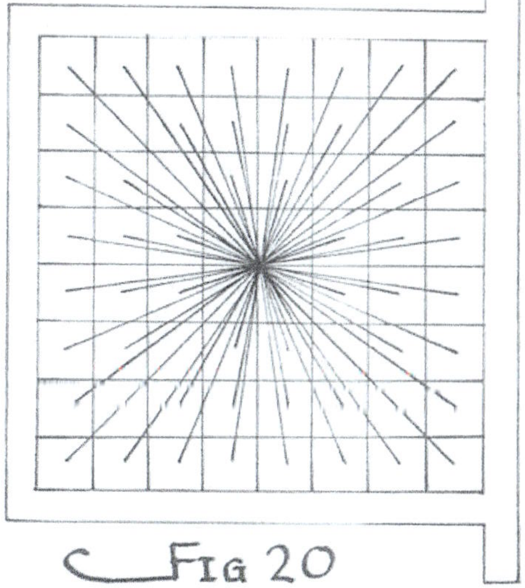

Fig 20

Page 203

ANSWERS

"Keep on remembering that the Universe is your own grandeur" (Swami Muktananda).

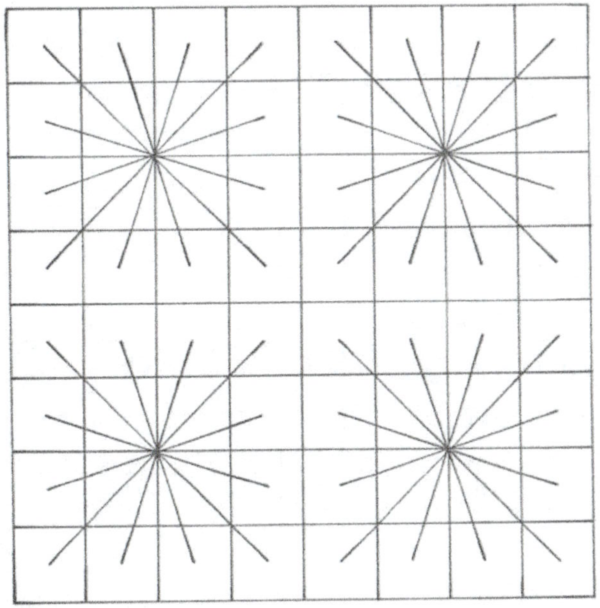

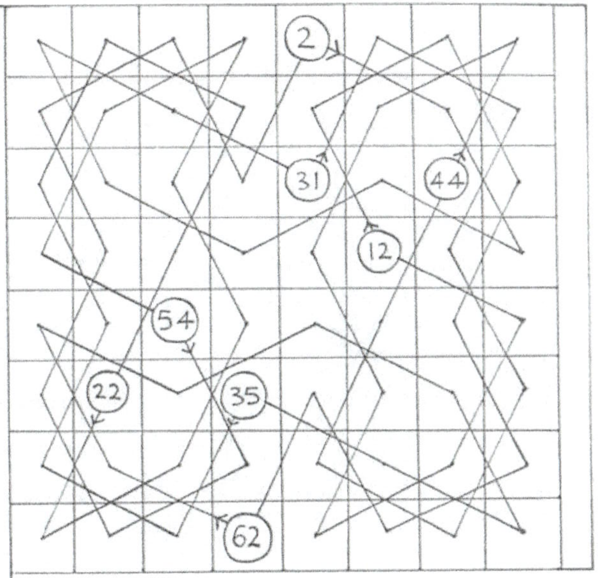

Fig 22

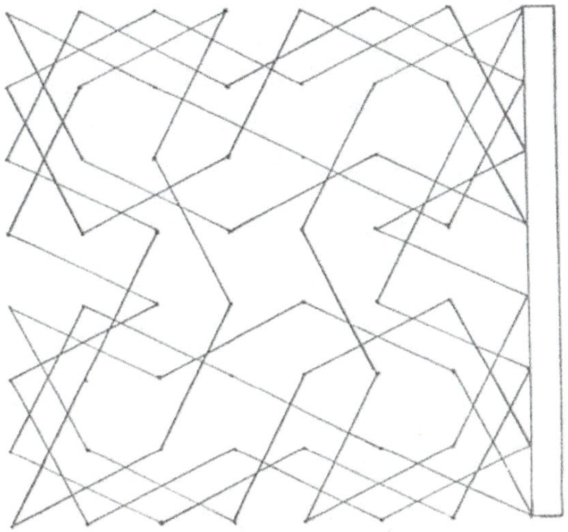

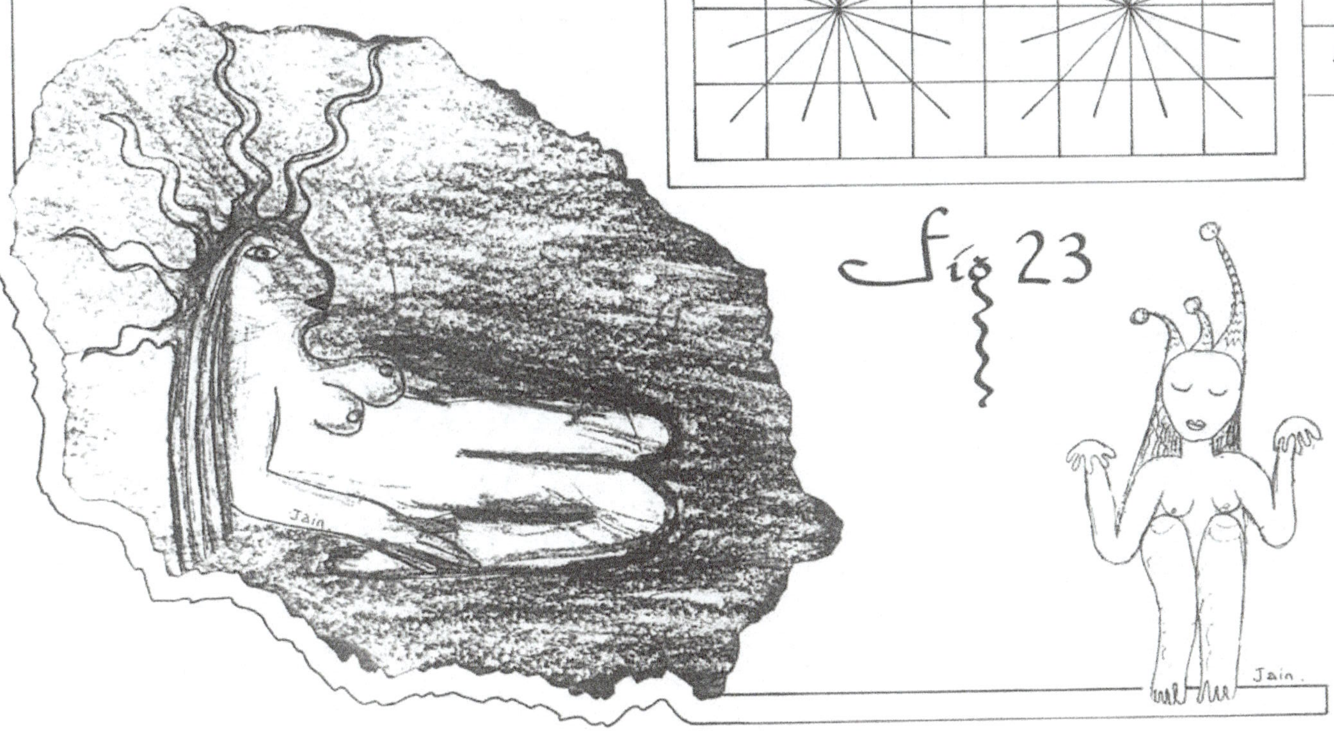

Fig 23

fig 27

fig 24

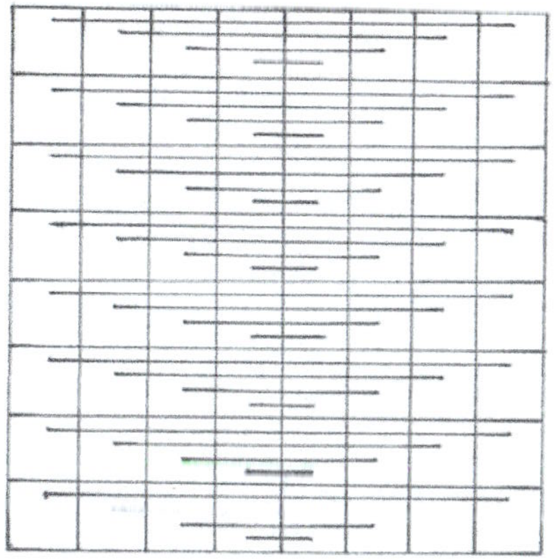

fig 28

Page 204

Page 205

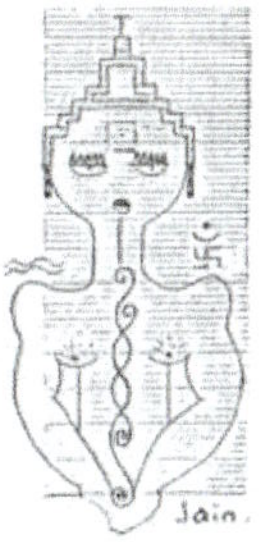

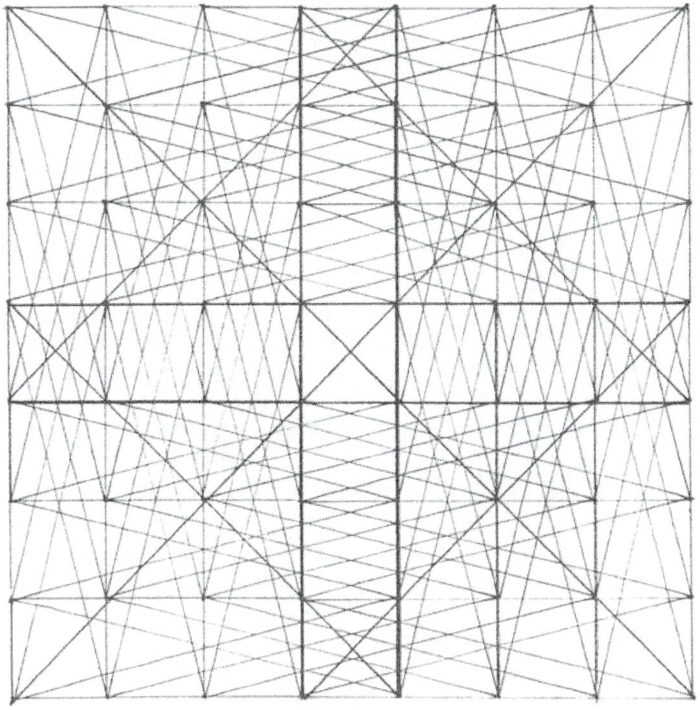

fig 31

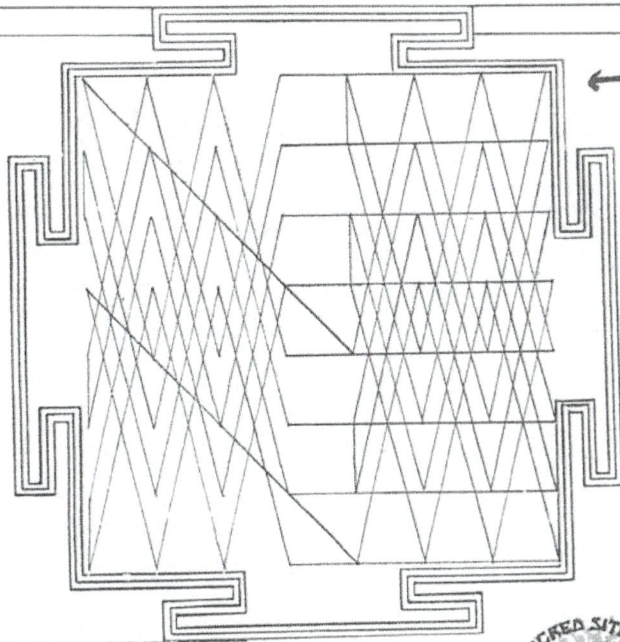

← fig 30

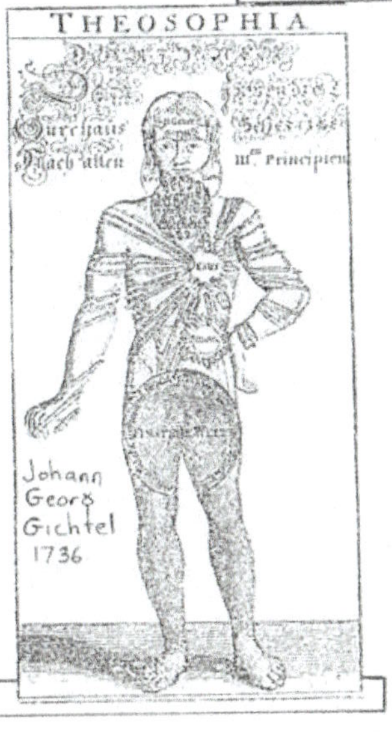

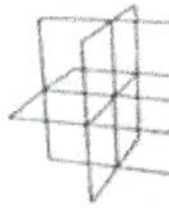

compendium, first published 1990,
is a F. R. E. E. D. O. M. S. publication:
For Research Expressing Essential Data Of Majik Squares!

C/O MAJIK SQUARE STUDIO
Box 396
P.O. BYRON BAY
2481
AUSTRALIA

I S B N 0 9594180 1 6

INTERNATIONAL ENCYCLOPEDIA
OF
HARMONY SQUARE
MAGIC

Page 206

Fig 34

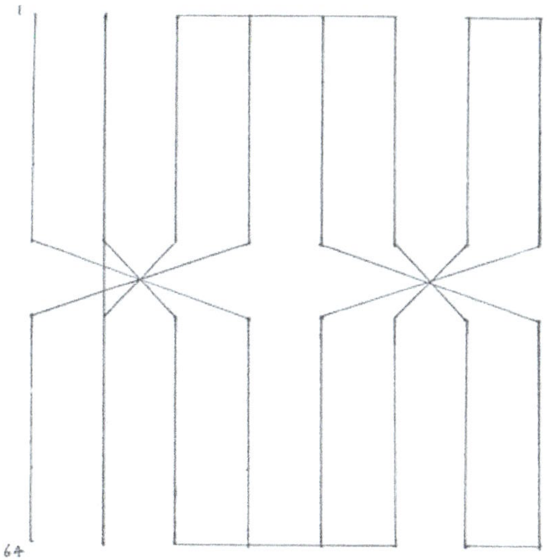

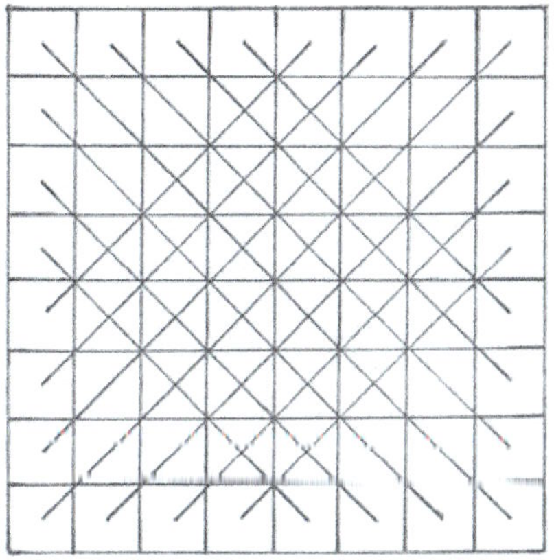

Fig 32

Page 207

CHAPTER NINE

SOLUTIONS

Fig 41

37	78	29	70	21	62	13	54	5
6	38	79	30	71	22	63	14	46
47	7	39	80	31	72	23	55	15
16	48	8	40	81	32	64	24	56
57	17	49	9	41	73	33	65	25
26	58	18	50	1	42	74	34	66
67	27	59	10	51	2	43	75	35
36	68	19	60	11	52	3	44	76
77	28	69	20	61	12	53	4	45

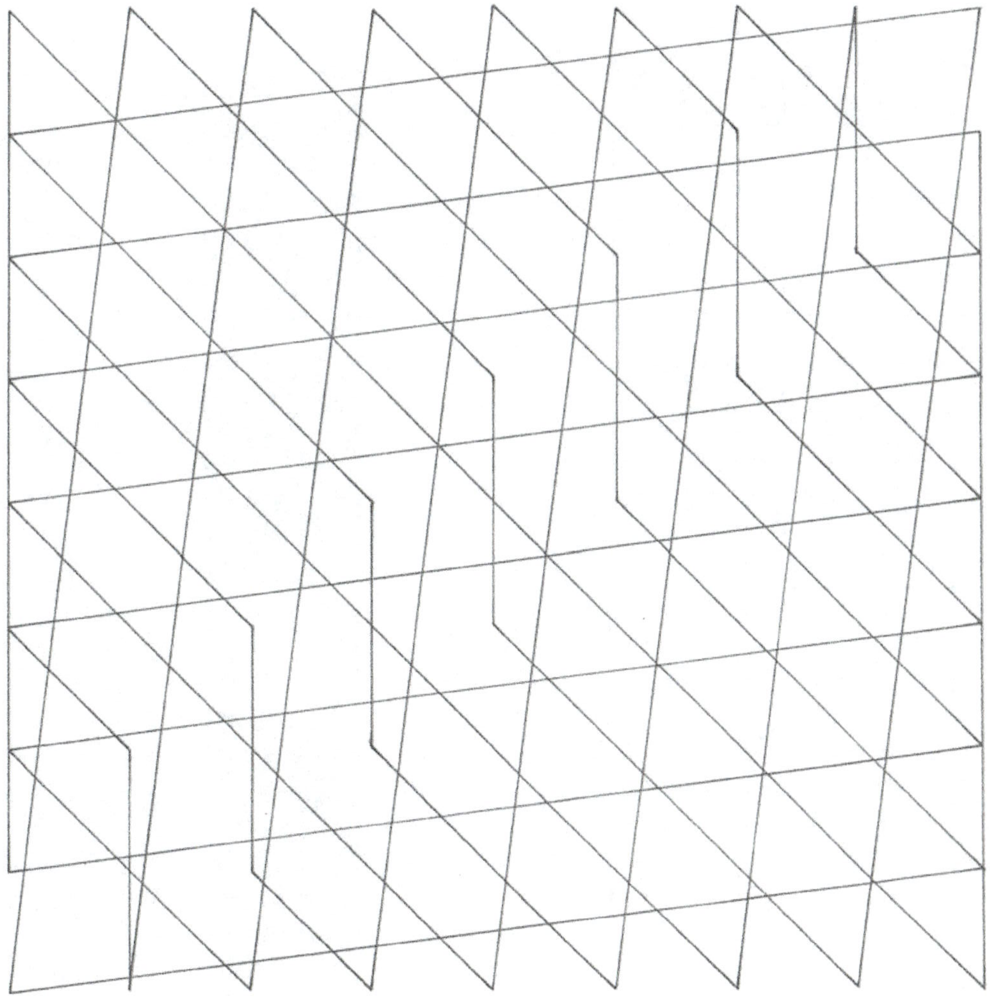

Fig 42

fig 44
Page 208
fig 44

Page 209
"Nothing is, that moves you,
You yourSelf are the wheel
which turns from within
And has no rest".
(a verse from :
Angelus Silesus).
ONLY WITH LOVE
PUBLICATIONS
"The Cosmic
Manifestation"
OF THE
MAGIC SQUARE
dynasty
Jain.
fig 47

38	6	16	33	43	11	28
46	14	24	41	2	19	29
5	15	32	49	10	27	37
13	23	40	1	18	35	45
21	31	48	9	26	36	4
22	39	7	17	34	44	12
30	47	8	25	42	3	20

Page 210 Fig 53

Shown above is the Essene Majik Square of 7×7 at 8 × 22.5°. Hand Drawn and just as accurate as a computer graphic, this Yantram is one 6 new m. Sq. Decals that may be silkscreened & available by 1996. To order the two existing Decal designs send $10 to Jain c/o P.O. Box '396' Byron Bay 2481. The two designs are a) m. Sq. of 4×4 & b) B. Franklin m. Sq of 8×8 shown here as Fig 3. Diameters = 135 mm. For a large size, diameter = 300 mm, of the Jupiterean 4×4, send $20. Costs cover PostaGe anYwhere in Australia.

Page 211

FIG 55

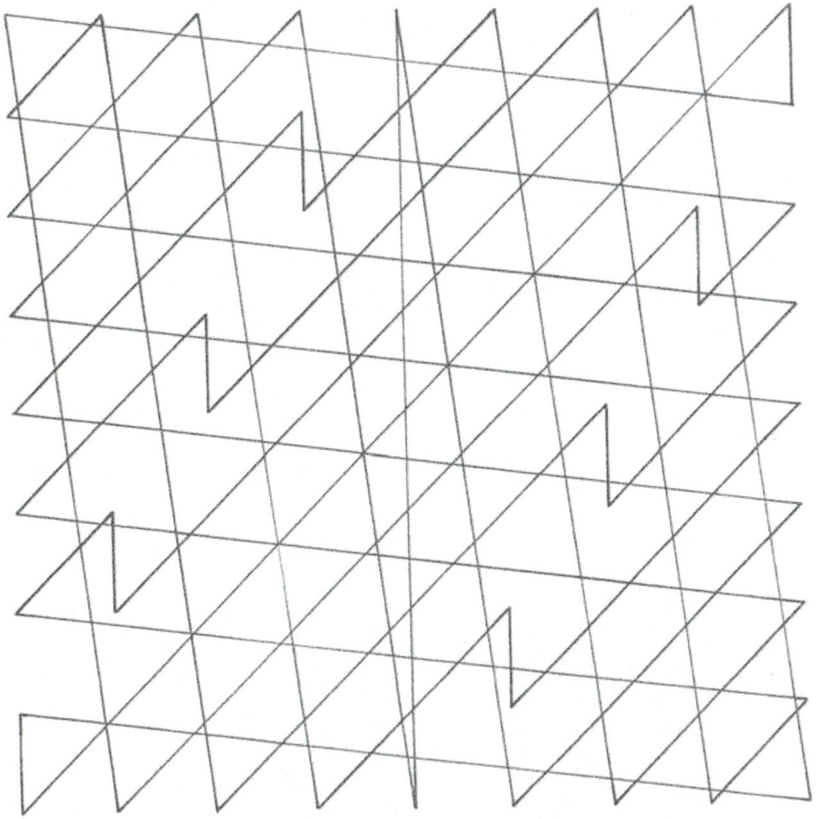

FIG 57

Fig 60

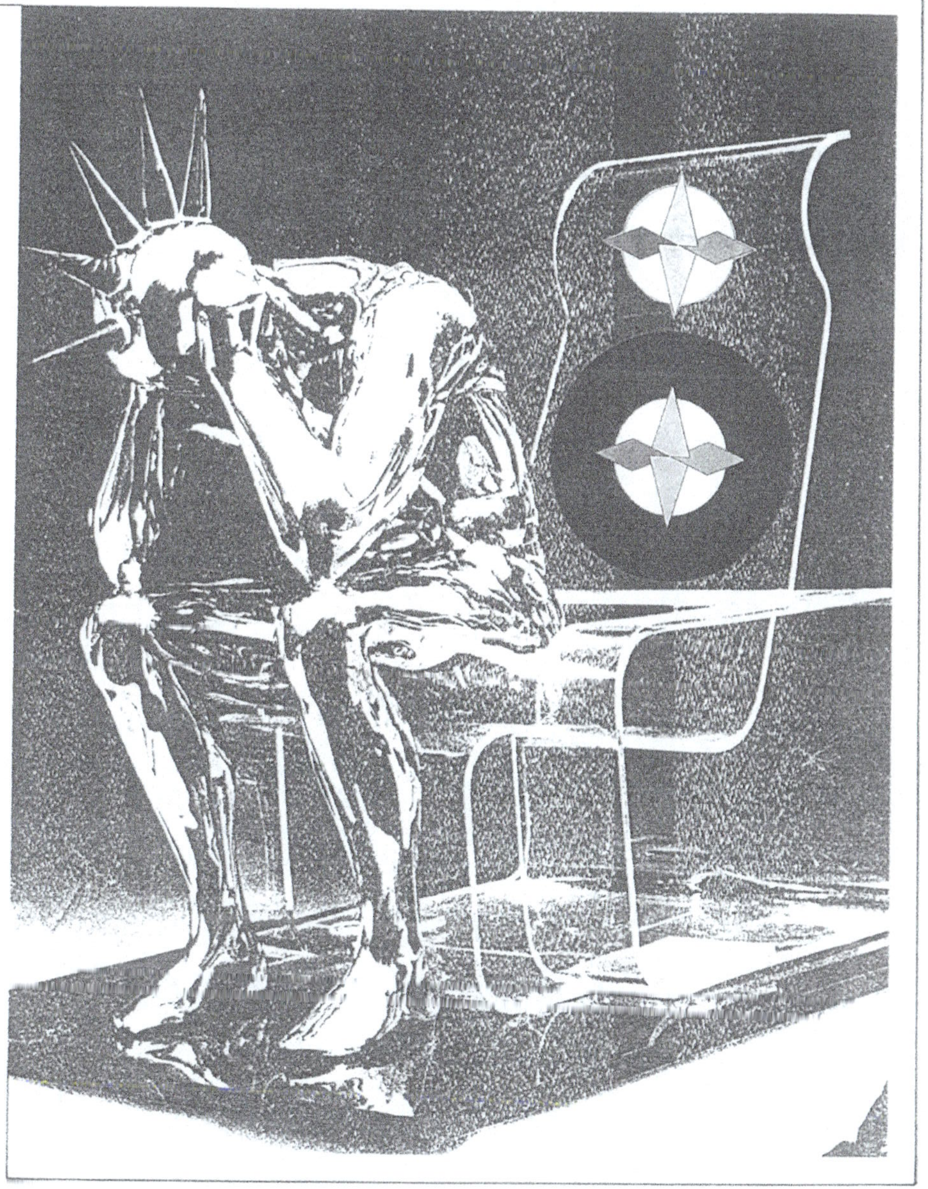

fig 62

FIG 64

"I do not know whether I was then a man dreaming I was a butterfly, or whether I am now a butterfly dreaming I am a man".

— Chuang Tzu.

(369-286a.c. Chinese philosopher & teacher).

Page 213

Page 214

Nijinsky in his costume for Les Orientales 1910.

"I AM unshakeable Self-Reliance & translucent Self-Awareness. I have no fear, dread or uncertainty". (Elizabeth Haich).

Border in temple form · Academicism, Munich 1839

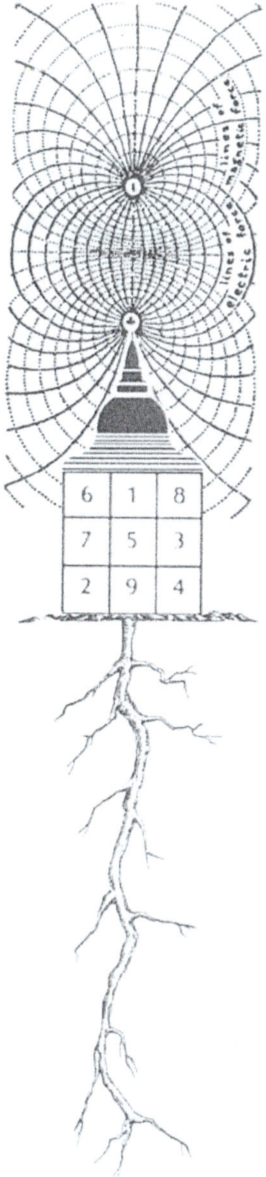

Field diagram of electrical polarities.

Transfiguring Affirmation.

(By Jesus The Christ)

The Ultimate Ascension Affirmation!

I AM THAT I AM

I AM the open door which no man can shut,
I AM the Light which lightet every man that cometh into the world,
I AM the Way,
I AM the Truth,
I AM the Life,
I AM the Resurrection,
I AM the Ascension in the Light;
I AM the fulfilment of all my needs and requirements of the hour,
I AM abundant supply poured out upon all life,
I AM perfect sight and hearing,
I AM the manifest perfection of being,
I AM the illimitable Light of God made manifest every-where.

I AM a son of God.

fig 82
fig 93
Jain.
Page 215
fig 94

Page 216 SOLUTIONS CHAPTER TEN.

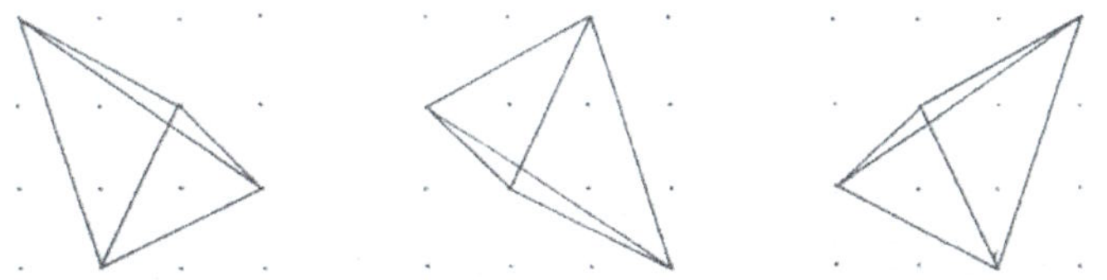

fig 97a fig 97b fig 97c

fig 98 at 0°

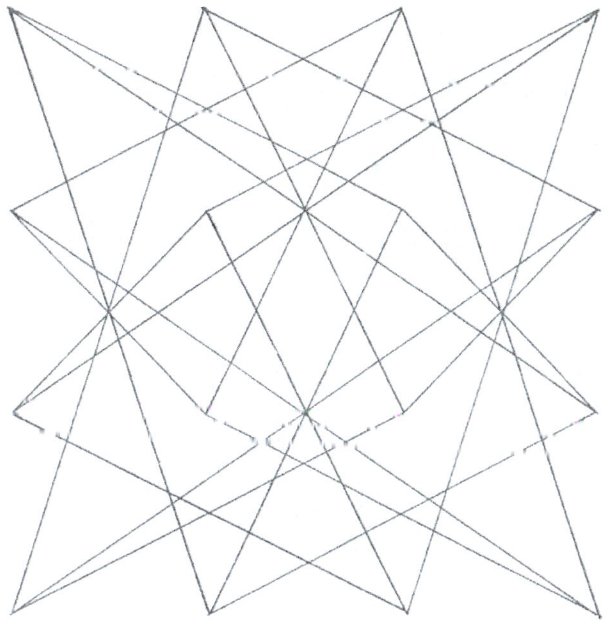

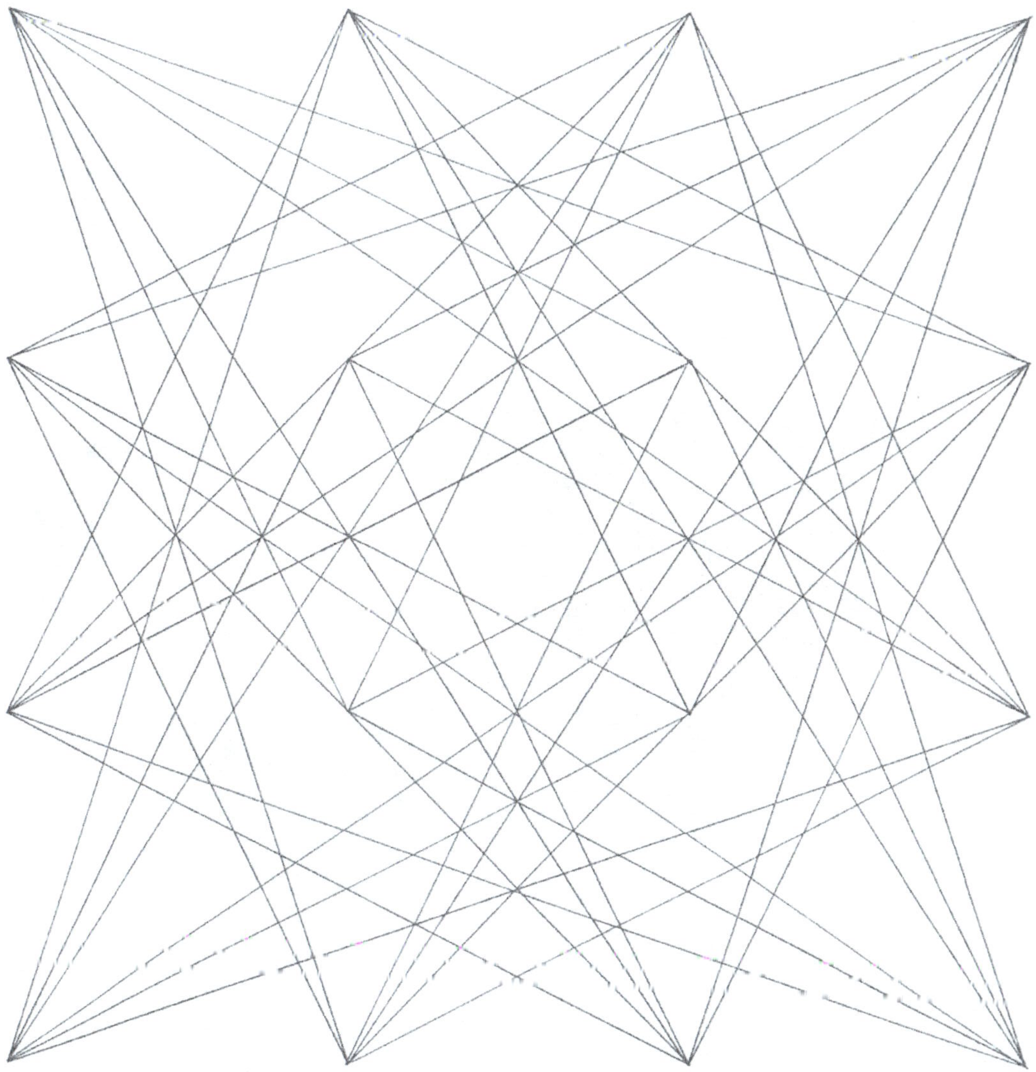

FIG 98 at 0°+90°

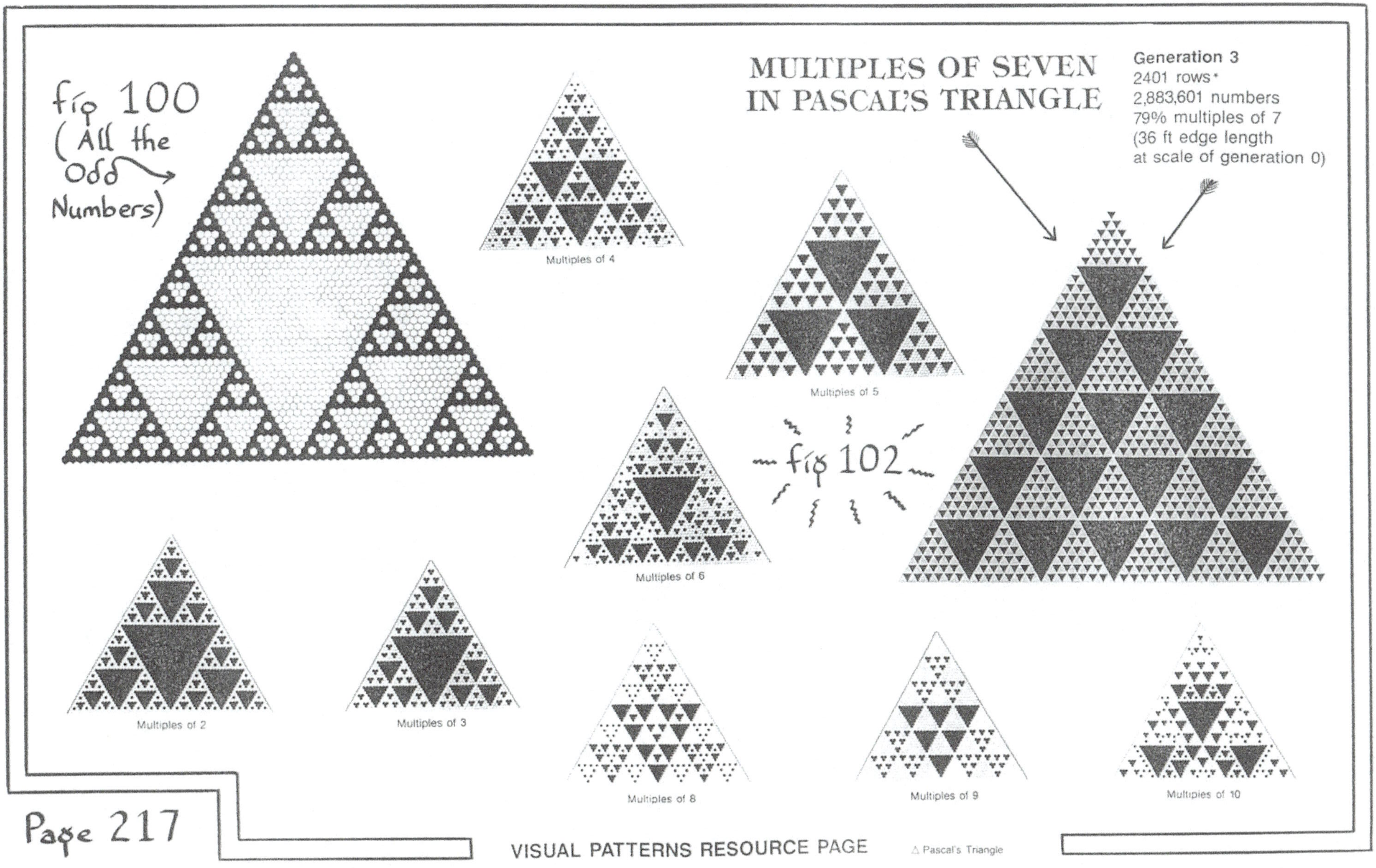

fig 100 (All the Odd Numbers)
Multiples of 4
MULTIPLES OF SEVEN IN PASCAL'S TRIANGLE
Generation 3
2401 rows*
2,883,601 numbers
79% multiples of 7
(36 ft edge length
at scale of generation 0)
Multiples of 5
fig 102
Multiples of 6
Multiples of 2
Multiples of 3
Multiples of 8
Multiples of 9
Multiples of 10
Page 217
VISUAL PATTERNS RESOURCE PAGE
△ Pascal's Triangle

ACKNOWLEDGEMENTS OF GRAPHICS USED + RANDOM NOTES

LEGEND : LHS = Left Hand Side
RHS = Right Hand Side
BOT = Bottom
CENT = Centre
M. Sq. = majik Square.
TOP = Top
P = Page

Acknowledgements

INTRODUCTION

- On This Page : LHS Drawing by Jain 1982, Torres Straits.
- P 1 a) Central Yantra is the Lo-Shu at 4×45° used as the front cover of my former book : "The Book of Harmony Squares" Vol. 1 1990. b) Angels on either side by Jain.
- P 2 a) Photo dedication to Asha Jain Sun, my second daughter né 27·12·1990, with almost 7 planets in Capricorn. b) Calligraphy of "T" + many other ornate initials are from a rare 1828 Bible in my Library. c) Bot LHS from a manuscript in the British museum C^th 14.
- P 5 Top RHS Dalai Lama as boy-king, Tibet.
- P 8 a) Drawing by Jain. b) Top LHS All opposing pairs sum 10 & there are 4 summations of 15 containing 3 cells each. Source Unknown. c) Top LHS circular border is the publisher's logo for "The Popular Encyclopedia".
- P 9 a) Border design is the M.Sq. of 3 tiled at 0°, 90°, 0°, 90° etc Refer to P 56 of Vol 1 which shows the hand-drawn version. This computer work was done by Nic Roche who helped me generate the first computerized M.Sq. graphics in 1991, Byron Bay studio. b) This drawing shows mySelf reaching out to connect 2 independent vines which I did join as a spontaneous & joyful ritual on an evening of Feb 1984. I had died on 27·1·84 due to a 6" knife wound thru my chest/heart region. After rehabilitating in the Blue Mtns N.S.W. I decided to visit my folks in Kirrawee (Aboriginal word for 'A Place of Flowers'). Behold the splendour of those flowers drooping over an arch of steel mesh at the entrance to the house. Thus symbolically, I connected all Polarities within me, reEarthinG & CentrinG mYself.

Page 237

Page 238

ACKNOWLEDGEMENTS : CHAPTER 8

ACKNOWLEDGEMENTS of GRAPHICS USED & NOTES, cont...

•On this Page. Bot RHS Drawing by Jain & Ali Kennedy 1994.
• **P 10 a)** Head of Pharaoh, probably Thutmosis III of 1st Dynasty. **b)** Top RHS Australian Optometrical Association symbol. **c)** Top LHS from "The Secret Teachings of All Ages" by Manly. P. Hall • **P 11)** bot LHS. M.Sq. of 8 Artform by José Arquëlles' "Earth Ascending" 1984. We are independently tuning into similar artforms or yantras or creations. It was a total delight to see this graphic validating my Life's works.
• **P 12** Top LHS from W.S. Andrew's "Magic Squares & Cubes" 1917. Dover. This book presented the pedantic aspect of M.Sqs & spawned for me 1000 original designs by transforming the M.Sq. boxes into visual artforms.
• **P13 a)** Top RHS Drawing by Jain. **b)** Bot RHS The 3 whirling swaztika-like segments from Tibet, known as rgyan-'k'yil. It often appears at the centre of the Indian Wheel of the Law, the Sansara, or ceaseless change or becoming.
• **P 14 a)** LHS Sankara arcarya, incarnation of Lord Shiva, powerful preacher of the Māyāvadi (impersonal) school of philosophy. **b)** Bot RHS. Hebrew script from "Transcendental Magic" by E. Levi.
• **P15 a)** Bot LHS from "Science of Soul" (Atma-Vijnana) by Shree 108 Swami Vyas Dev Ji Maharaji published by "Yoga Niketan Trust" Bharat (India) 1964 **b)** Top LHS symbol was found in a booklet "Practical Holography" a research company based in Tweed Heads N.S.W. Its interesting to note that it is identical to the top of the Franklin M.Sq. of 8x8 Yantra (Fig 3). I've been disseminating these M.Sq. Graphics for a Decade now & they appear in many forms in many books, posters.
• **P16** Drawing by Jain. • **P 17** LHS from "Crystal Visionary" mag, compiled by M. Ney.
• **P19 a)** Top LHS Angels by Jain. Originally handpainted upon a 3m x 3m M.Sq. Banner/Mural **b)** Bot LHS. Paul Carus' quote from "Magic Squares & Cubes" by W.S. Andrews.
• **P 21** Priesterweihe durch Papst Johannes Paul II. im Petersdam. Von Ängslen und Zwängen verformte Menschen. • **P 22** Buki designs, Mandala Color Pad 1979.
• **P23** RHS Drawing by Jain • **P 24** Bot RHS Drawing by Jain. Above the holy man's head is a mandala aura which is the M.Sq. of 4x4 at 4x45°, drawn on a computer.
• **P 26** RHS Drawing by Jain for my daughter Asha 2½ weeks young on 15/1/1991 Wategos Beach Byron Bay, when 7 planets aligned in Capricorn. • **P28** Bot LHS Drawing by Jain. (derived from a M.Sq. of 8x8). It shows a typical triangular form used in many of my early works. It is a Golden Triangle whose proportions are 1:1.618 as found in DNA, Pyramids, M.Sq. Artforms. This figure has come to symbolize "HAHA-AHA" my Spiritual teacher of the Sacred Geometries. Originally, this book was to be called "The Book of Magic Squares" thru HAHA-AHA (Haha = Sound of Laughter, Aha = Divine Realization, Thibet)
• **P 29** LHS Magician from front cover of a Crystalight mag. called "Wizardcraft". The Artist & Source are unknown, as are many images in this compendium. Collaged at Top LHS + the 6-pointed Spiral Star are symbols of Infinity in Space & Time, emblem of our Creators, Elohim. Raël Movement. • **P 31** Vaisnava ācārya, follower of Lord Vishnu who teaches by his example. • **P 32 a)** The Jains were a fanactic Buddhist sect who killed nothing & owned nothing, they were the naked Indian worshippers who constructed M.Sqs. for intellectual past-time. "Jain" was my name for all my Art + Geometries, for years, without knowing that they were an Indian sect, Let alone beinG the ancient Masters of the "Jaina MaJik SQuares".

"The Prayer for the New Age"

I am the creator of the universe
I am the father & mother of the universe.
Everything came from me.
Everything shall return to me.
Mind, spirit & body are my temples,
for the Self to realize in them my Supreme Being & Becoming.
(1988 Maitreya).

• P 32 b) This Jaina Sq. (Brahmin's Knight's Tour) was presented by Edward Falkener 1892 in "Games Ancient & Oriental". Published by Longmans Green & Co London, NY 1892. c) Bot RHS ½ horse, ½ human Indian image depicts Ashvins.
• P 33 "The Women inside the Horse is not meant to depict male domination over women, but rather that the symbolic Knight's tour is really a search for his Shakti or Female Essence, so as to become One. Both images from "Erotic Postcards". Top horse called "Le Vainqueur", the other "white Stallion".
• P 34 Bot RHS image from "Toward the One" Hazrat Inayat Khan 1974 Sufi founder.
• P 35 a) Top LHS heart drawing by Jain. b) Mid LHS Heart Dolphins, logo for "Shettim" festival, Byron Bay, 1992. c) Top RHS ngo forum on women, Beijing 1995.
• P 36 Garuda is Lord Vishnu's eagle carrier.
• P 37 a) Excerpt from "Guiding Manual for a New World" by Jean Hudon. World Federation for Peace. b) RHS drawing by Jain.
• P 38 a) Bot. Cent. Fig 24 is from "More Mathematical Activities" by Brian Bolt. • b) Drawing by Jain.
• P 39 a) Bot LHS design by Keith Critchlow, cosmic Architech & a Prof of Majik Squareology & ·618–ology.
b) Top LHS computer graphic photo by Barry Blackman.
• P 40 a) Cent. reprint from "You Can Heal Your Life" by Louise Hay. b) LHS Angel by Renée Garling.
• P 41 Top LHS stylized angel emanating from lotus, is a logo by Tia Demou. Byron Bay.
• P 43 LHS typical Indian 'Sadhu' wanderer. His trident is the Shiva symbol.
• P 44 from top then LHS: Satyananda, Holy Mother Amritanandamayi (largest face), Hazrat Inayat Khan, Yogi Srimat Kuladanda Brahmachary Ayananda, a Barber woman & Maharishi Mahesh Yogi.

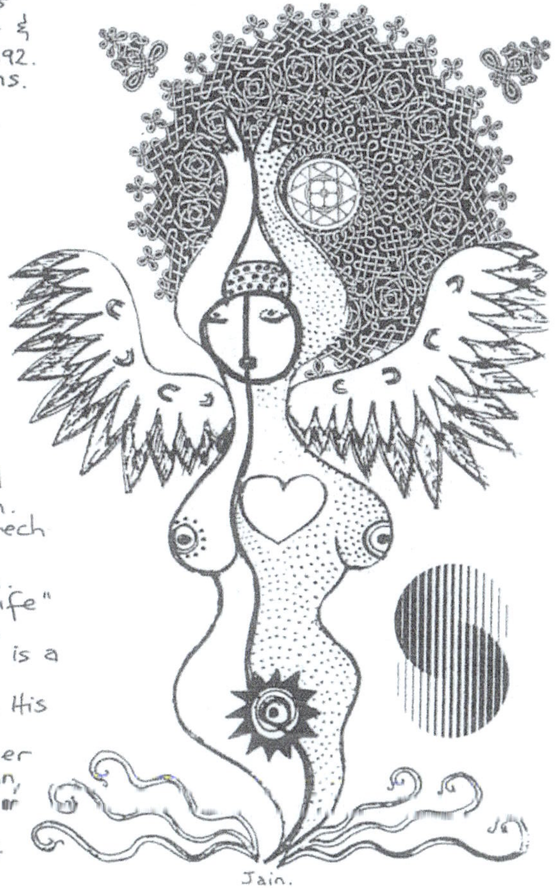

• P 45 Icon of Mahavira. The human form is shown as an empty space to represent the Formless, the Spirit. • P 46 Lord Krishna, Incarnation of Beauty.
• P 47 a) LHS Spirograph border b) Bot Cent. Narayan & his consort Laksmi.
• P 48 a) "FunDesign" c/o Mondragon Industries b) Bot. Cent. Drawing by Jain.
• P 49 Drawing by Jain. The symbols collaged are Syske Bank Jutland, Junghans Funkuhren, Schleicher & Co, Scottish Development Agency. The idea is to immerse buisness companies, by use of their psychic logo, into an loving & intelligent Field of Majik Square Harmonics to invoke honest & humanitarian dealings.
• P 50 a) RHS Yantra from "Living On the Earth" by Alicia bay Laurel. b) bot LHS Drawing by Jain c) Top LHS hand-drawn Yantra by Jain of the M.Sq. of 7x7 at 8 x 22·5° originally designed for a series of colored Decals in 1995. WE ALWAYS MANIFEST DIVINITY.

Page 239

• P51 a) text adapted from "Yantra" by Madhu Khanna (Tantric Symbol of Cosmic Unity).
b) Bot RHS portable 8-sided marquee, photo courtesy of the British Museum c) 2 Drawings by Jain in centre, over text.
• P52 Magic Rotating Ring from "Magic Symbols" by F. Goodman • P53 a) Bot LHS Circle from "Form Drawing" by Niederhäuser + Frohlich (Steiner-based) b) the circle adjacent to that is an Islamic Pattern by Paul Marchant 1984.
c) Far LHS Drawing by Jain d) Drawing in centre by Jain • P55 Spiral collages on RHS are known as Baravalle Spirals.
• P56 Bot RHS "Moses" by Michael Angelo.
• P57 Drawing by Jain 1984 which became a large 3m x 3m mural at the entrance to Arthaven Community in the Blue Mountains, N.S.W.

ACKNOWLEDGEMENTS : CHAPTER 9

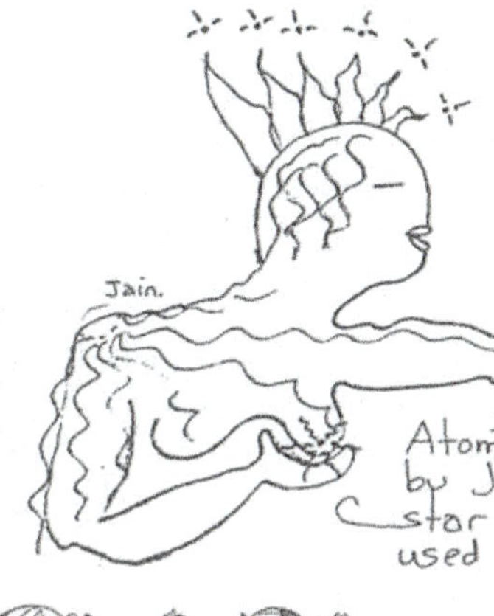

• P58 Border is a tiling of the Aust'n Optometrical Ass'n symbol
• P59 RHS door-sized mural of a series of E.T.s, Jain, 1984
• P60 a) Top LHS Arabian tiled 9-pointed star motif
b) Bot. Cent. emblem of Tibet: The Eternal Knot.
c) Top RHS Krishna in his Atomic form by Jain • P61 a) Drawing b) Top 9-pointed star symbol of the Bahai Faith used in ground plans of temples.

• P62 Cent. Drawing by Jain.
• P63 LHS Arun Lady, from Hanu, Leh City, Ladakh (geographically Tibet in far Nth India).
• P65 a) Poster of Laxmi, Goddess of Wealth, re-collaged by Jain. Artist M.C. Jeganath. Pon. Pudupatti.
b) circle beneath Laxmi's head is Leonardo da Vinci's knot. • P67 LHS Photo from "Australian Photography 1970's mag.
• P68 a) LHS text from "An Intro to The KEYS of ENOCH" by J.J. Hurtak. b) Bot Cent. Drawing by Evelyn Krosby, Blue Mountains 1983
• P69 Cent. I call this drawing "5-Steps towards Cosmic Consciousness" 1982 Cedar Bay, Full Moon ♓ copied from a science mag. Jain.
• P70 LHS front cover for Omna's book "In The Chalice Of The Heart" Delhi 1989. Concept by Omna, Drawing by Jain. • P71 The 9th & highest heaven of Dante's Divine Comedy. The geography of heaven as a series of spheres reflecting the circuits of the planets is a constant image over 1000s of years, transcends cultural boundaries.
• P72 a) LHS meditator from "Atma-Vijnana"
b) Bot RHS sketch by Kahlil Gibran
c) Bot. Cent. Drawing by Jain.

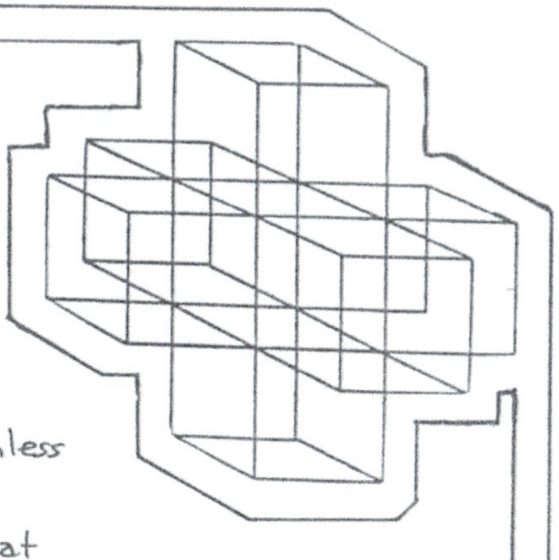

On This Page a) LHS Drawing by Jain b) Bot RHS from Kircher's "Ars Magna Sciendi" regarding the problem of Diversity, Kircher arranges 18 objects in 2 vertical columns & then determines the number of arrangements in which they can be combined. Conclusion : infinite individuality & variety of Being. Life is rich.

• P 74 a) "The Gates of Dawn" by C. Yandell & N. Walker. b) LHS Circle over lap is a design by Paul Bonet from a book cover, 1st edition of Dionysius Hilicarnaseus (1480).

• P 75 Bot RHS "Genuine Turkish Delight" from a book "Erotic Postcards"

• P 76 RHS Drawing by Jain. • P 78 Bot. Cent. Drawing by Jain.

• P 79 Cent. Soul Bird drawing by a friend/lover Julia, Sydney, given to me as a gift, in 1978. • P 80 Top RHS. drawing by Jain.

• P 82 a) RHS Drawing + words thru Shahastra from "Invisible Worlds" b) top RHS circular logo is the spiritual symbol of BabaJi the Hymalayan Deathless master. • P 83 LHS Drawing by Jain.

• P 84 RHS from a Buddhist Thai mag. front cover of intertwining dragons.

• P 86 LHS Drawing by Jain 1984, one of my favourites, of the 1000s that I have penned, this particular one says a lot.

• P 87 a) Great Sphinx of Giza, quardian of the dead, has body of a lion (♌) & head of the pharaoh Kahre, near whose pyramid it crouches. 240' long. 66' high. We are really in the Age of Leo Cusp Virgo, this 2000 year span, not Age of ♒. Virgo = ½ woman face of the Sphinx. Many secret chambers lay beneath it! b) text extracted from "The Bridge to Infinity" by Bruce Cathie N.Z. This article first appeared to me in "Southern Crossings" mag Vol 5 No. 1. • P 88 a) Cent. text from "Windwords" mag. Vol 2 No. 12 June '88. Ramtha. b) Earth's Crystal Grid developed by Bruce Cathie, reprinted in "The Crystal Visionary" Vol 1 by Michael Ney 1986. • P 89 a) This Prayer is taken from my first self-published book "The Time Twins" by Enfant-de Neptunuu. 1981 p54. At that time, age 24 yrs my Lebanese name since birth was Collin Nicholas Saad (or C.N.S. or Central Nervous System) now a.k.a (also known as) Jain or Jaina. b) Bot. RHS Drawing by Jain c) RHS 16 frequency circle / dome pattern by Buckminster Fuller's computer graphics. • P 90 a) LHS Paul Bonet, great French designer. The 1st edition of "Dionysius Hilicarnaseus" 1480, is bound in his "irradiant" style in which apparently parallel lines, varied with great subtlety, create an impression of 3 dimensions. b) Top. Cent. presented by "The Spiral of Friends", A Fourth Way School organized by James Vincent Randazzo.

• P 91 Bot. Cent Drawing by Jain • P 94 a) Bot. RHS Rangoli pattern from "The World's Great Religions" Life books b) text of 1st paragraph inspired by "Insight" mag ("Into the Hindu experience") Nov '84 c) 2nd Paragraph inspired by "Yantra"

• P 95 Fig 73 is from "Yantra" by Madhu Khanna (Thames & Hudson).

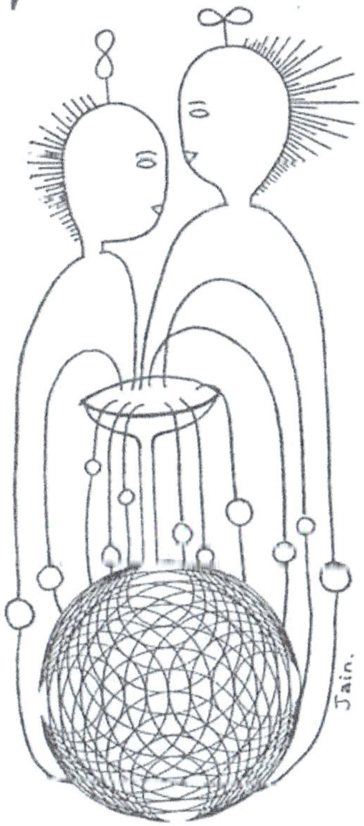

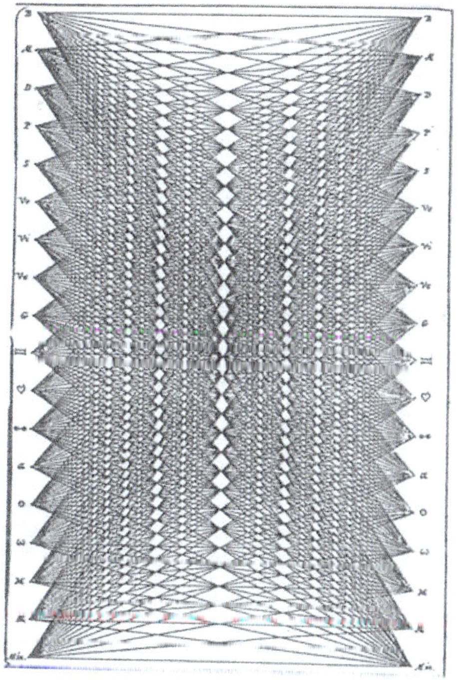

Page 241

Page 242

• P97 Holy Pushkar Lake, Rajasthan, the only Brahma temple in India. Performance of a Fire Ritual (Agni Yagya). The main sources of notes for this chapter on Sri Yantra + Fig 71 are from "Sri Chakra" by S.K. Ramachandra Rao 1989 + "Yantra" book.

• P98 RHS Yogi from "Science of Soul" (Atma Vijnana) by Vyas DeviJi. Artist un-acknowledged.

• P100 a) LHS from "Yantra" b) RHS text from mag. "Insight into the Hindu Experience".

• P101 a) LHS sanskrit m.Sq. Fig 78 from a book, all in sanskrit, translated as "Yantra Vidya" with 121 illustrations by Siddha Baba Hoddarnath. This book was given to me by an Indian Yogi "BapuJi" whose room was framed with 4 m.Sqs. one on each wall. He could bury himself underground for 2 days by slowing down his heart-beat to re-emerge alive. Many scientists worldwide have recorded him. I saw 100 newspaper clippings verifying this.
b) RHS Ardhanageshwari (½ Shiva, ½ Shakti) from an Indian poster by J. Khanna & Co. Madras.

• P102 a) text from "Investigations in Mathematics" by Lorraine Mottershead. b) Cent. Drawing by Platpus (a.k.a Radiance) from Wytaliba Community N.S.W my first tantric initiatress.

• P103 a) Song lyrics by Jain. b) Cent. Drawing by Ev. Krosby 1983.

ACKNOWLEDGEMENTS :

CHAPTER 10

• P104 a) BabaJi, mentioned in Parahamsa Yogananda's "Autobiography of a Yogi" published by Self-Realization Fellowship 1979, California.
b) from "Dossier on the Ascension" by Serapis Bey. This image I waited for for 15 years to determine how to construct the true 51°51' pyramid angles.

• P105 a) Bot RHS inside cover of "Srīmad Bhāgavatam" 2nd Canto. Part One depicts Sukadeva Gosvāmī wisely answering questions from King Parīksit.
b) Top Cent. An Indian Divine Mother Devotee possibly related to Sri Aurobindo.

• P106 Bot LHS Drawing by Jain 1982 inspired by a cartoon, artist unknown.

• P109 Bot RHS the whole area of the drawing is a Golden Mean Rectangle in which the central motif is repeated at each corner (from Keith Critchlow).

• P110 from "Planet Heartworks" by Dan Winter of Crystal Hill Farm, now California, 1988, a treatise on "the decagonal shape of the gene & the miracle of the quantum leap of ratio from cell to gland to emotion to planet shapes as frequency signatures. A Golden mean Path with Heart".

• P111 a) LHS Fig 84 is by Bruce Cathie, also author of Harmonic 33, developed a concise mapping of the Earth's ley-lines (that offer Humanity a free & eternal energy system). This was by observing the paths of UFOs & positions of sacred sites. See also pp 87, 88, 89 of this book you are reading now. Also, the Essene m.Sq. of 11 (Fig 111) is very similar to another of Bruce Cathie's Polar Grid designs. b) top RHS & Bot. RHS buddhas are images from "The Power of Limits" by György Doczi 1981 Shambhala c) "The Ship at Sea" by Jain 1982.

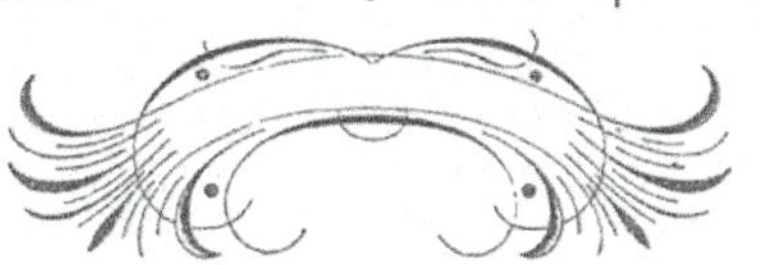

CREDO

of the International Biogenic Society

We believe that our most precious possession is Life.

We believe we shall mobilize all the forces of Life against the forces of death.

We believe that mutual understanding leads toward mutual cooperation; that mutual cooperation leads toward Peace; and that Peace is the only way of survival for mankind.

We believe that we shall preserve instead of waste our natural resources, which are the heritage of our children.

We believe that we shall avoid the pollution of our air, water, and soil, the basic preconditions of Life.

We believe we shall preserve the vegetation of our planet; the humble grass which came fifty million years ago, and the majestic trees which came twenty million years ago, to prepare our planet for mankind.

We believe we shall eat only fresh, natural, pure, whole foods, without chemicals and artificial processing.

We believe we shall live a simple, natural, creative life, absorbing all the sources of energy, harmony and knowledge, in and around us.

We believe that the improvement of life and mankind on our planet must start with individual efforts, as the whole depends on the atoms composing it.

We believe in the Fatherhood of God, the Motherhood of Nature, and the Brotherhood of Man.

—composed in Paris in 1928 by Romain Rolland and Edmond Bordeaux Szekely

- On this Page a) Top LHS a M.Sq. of 5x5 Yantram at 4x45° designed as a Decal 1995 for the international market, bearing the 9 numbers of the M.Sq. of 3.
b) Bot RHS : The Ascended Master EL Morya, Chief of the Darjeeling Council of the Great White Brotherhood.
- P 112 Top LHS Fig 90a & 90b are my personal contributions to this subject of Phi & M.Sqs. because it bridges the two subjects. There are Professors of Phi Studies & pedantic Magic Square Theories & you would think that the two subjects are unrelated, but I have shown that M.Sq. Artforms are full of these Golden Ratios, that they also relate to the onion-shelling of electronic orbits about their nucleus & that the centre of the Tibetan Calendar's magic square of 3x3 is identical to the phi-ratioed atomic structure of Beryllium. (Refer to Ch. 12 P 152)
- P 113 "My Ten Fingers" Drawing by Jain, my first attempt to draw my hands. India.
- P 114 fig 92 Drawing by Jain • P116 The Ten Puzzle first appeared to me in "Cole's Funny Picture Book" No. 3 p 189. • P118 a) Top Cent. Drawing by Jain. b) information + diagrams on pp 118 & 119 are from 2 sources "Pascal's Triangle" by T. Green & C. Hamburg, & "Visual Patterns in Pascal's Triangle" by Dale Seymour.
- P119 a) Top LHS Sun from Robert Fludd material C16th b) Top RHS Angel by Jain c) Cent. Man in Spiral from a Claude Bragdon book.
- P 120 The use of the word "hir" is borrowed from Timothy Leary's balanced gendered writings where "him or her" = "hir" & "He or She" = SHe.
- P121 Top LHS + Top RHS from Manly. P. Hall's "Secret Teachings of All Ages" • P 122 a) Top LHS Star from "Form Drawing" by Niederhäuser + Frohlich. b) LHS large circle – front cover of "The Spiral Dance" by Starhawk. c) RHS diagrams + fig 107 from "Sacred Geometry" by Robert Lawlor. • P 123 Top LHS from "The Round Art".

ACKNOWLEDGEMENTS : CHAPTER 11

- P 124 This M.Sq. of 11 Yantra, joined side by side, was first published in "The Time Twins" by Jain 1981 (a.k.a Enfant-de Neptunuu.
- P 126 a) All the text is from the Editor's Brief Commentary from "The Gospel of the Holy Twelve" (This is a rare book in my M.Sq. Library as I have not seen another like it in 15 years. b) LHS Drawing by Jain.
- P 127 a) LHS Drawing by Jain. b) Top RHS & Bot. RHS by Kahlil Gibran whose quaint home I visited in the Lebanon Snowy Mountains. Being myself of Lebanese descent, I quite relate to this artist/author who said things like "Your Children are not your Children..." etc.
- P128 Top Cent. Drawing by Jain. • P129 LHS Drawing by Jain.
- P 130 Top RHS M.Sq. 11 Yantram fotocopied from a photo of a 2m x 2m multi-colored cotton mural by Evelyn Krosby 1985 Kippaxs Studio N.S.W.
- P 131 Bot LHS Horse from an instructional Sumi-e book that I once studied from, title unknown. Art with meditation & with one pure stroke.

Page 243

1 of 2 APPENDIX

REQUIREMENTS

- A large Hall, Library or Double Room. This will allow 4, 5 or 6 large 3m x 3m Magic Square Paintings (cotton mural tapestries) to be hung. This gives the feeling that the students are entering an Art Gallery
- A minimum of 50 students per class per 2 hour session
- 2 classes per day. One morning and one afternoon session.

TOOLS NEEDED

Each student will be required to bring in:

- Coloured Pencils
- Lead Pencils
- 300mm Ruler
- Sharpener
- Rubber
- If there are not enough desks, students can use kneeboards.
- The teacher can supply a door-sized Pin-Board with 2 boxes of thumb-tacks. This allows us to pin up and view all the geometric drawings, tessellated side-by-side, for the whole class and other teachers to view.

6	1	8
7	5	3
2	9	4

p154

HOW I TEACH

When teaching MAGIC SQUARES I create the room as a theatre. Walls are draped with colourful Magic Square Murals the size of 3m x 3m. I wear a special Merlin-like robe that depicts the lesson. Upon a large stand, I have what is called a **'Human Cloth Book'** which has 30 cloth pages (1200mm x 1200mm) of Magic Square theory and design. It is a unique hand-sewn garment that can be worn and is used throughout the whole show.

One of the main lessons shows how the common Multiplication Table produces the Atomic Structure of the Platinum Crystal!

ACROSS THE K.L.A.S

What I offer is creative and Multi-Cultural Mathematics whose roots stem from India, Egypt, Arabia etc. I relate how Mathematics can be inter-subjectal or crosses Key Learning Areas:
i.e: Maths as Art, Maths as Science Maths as History. e.g. **The Magic Square of 3 x 3 is the centre of the Tibetan Calendar !**

Also, Arab's domed mosques used Magic Squares in a secretive and elaborate way. Benjamin Franklin was a closet genius on this subject.

COSTS

- Each student will be requested to donate $4 for a 2 hour session.
- All travelling expenses.
- A minimum of 2 Resource Books (Teacher's Manuals) are required to be purchased for the School or Maths Library.

MATHEMAGICS

ATTENTION: ALL SCHOOLS

MAGIC SQUARES,
SACRED GEOMETRY
&
VEDIC
MATHEMATICS
SEMINARS
for School Children
of all ages.

*This **Maths Enrichment Program** is the only registered travelling show in Australia that caters for creative Mathematical exploration*

Presented by **JAIN**

2002

BIOGRAPHY

I have been invited to lecture at several Maths Conferences in Sydney and Universities in North NSW. The last 12 years have been devoted to School Performances with Mathemagics in High Schools and Primary Schools all over Australia.

Gifted and Talented Seminar organisers employ me to teach at their Maths Camps. These are Carnivals where 20 or 30 schools may be attending.

Private, Independent, Catholic, Steiner and Montessori Schools also employ me to rekindle the beauty of Mathematics. The last 20 years have been devoted to the collection and theatrical education of Magic Squares, Magic Cubes and The Golden Section (PHI) symbolised by ϕ.

This year, I have made 2 important contributions to the mathematical world:

1): **I discovered a distinct pattern in the PHI Ratio** (1 : 1.618) which for thousands of years was considered infinite and without pattern. Phi Decoded will be released and documented in mathematical papers. It is recorded, for the first time in print, in my latest book: **THE BOOK OF PHI.** 2): A new and more intelligent formula for the entity known as **e = 2.71818**, another transcendental number, has been identified.

The Institute or Archives so far established for the promotion of mathematical discovery is known as '<u>F.R.E.E.D.O.M.S.</u>' which is an acronym for: 'For Research Expressing Essential Data Of Magic Squares'. I currently run Seminars on Sacred Geometry and Vedic Mathematics for adults and children.

2 of 2 APPENDIX

BOOKS BY JAIN

- **The Book Of Magic Squares: Vol. 1,** pub. 1990. – 164 pages –
- **The Book Of Magic Squares: Vol. 2,** pub. 1996. – 160 pages –
- **The Book Of Magic Squares: Vol. 3** pub. 2000. – 174 pages–
- **Join-The-Dots,** **The Art of Magic Squares, Book 2,** pub. 1999. – 56 pages –
- **The Magic Of 9 In Vedic Mathematics** pub. 2001. – 124 pages –
- **The Vedic Mathematics Curriculum For The Global School** pub. 2001. – 142 pages –
- **The Book Of Phi,** **The Living Mathematics of Nature** pub. 2002.

VIDEOS

- **Vedic Mathematics For The New Millennium.**
  Filmed: 2002. Byron Bay
  Duration: 2 hours 15 mins.
  16 Sutras for Mental, One-Line Arithmetic. How to become a Human Bio-Calculator. Improves Memory.

0155

ALL ENQUIRIES:

JAIN

c/o MAGIC SQUARE STUDIO
777 Left Bank Road
Mullumbimby Creek,
NSW 2482.

Phone:
0266 844409

The Ultimate Majik Square.

Page 287

The **CONSTANT = 0** i.e: The Sums of the 3 columns, rows + 2 diagonals all equal 0.

p156

Jain Products Order Form

NEW! **The Living Mathematics of Nature, Series – 5 DVD Set, pub: 2006**

	Item	Price	Weight
☐	**Complete 5 DVD Set** (Special! Save $25)	$200	240 gm
☐	**Introduction to Ancient Knowledge** (DVD #1 of 5)	$ 45	100 gm
☐	**Vedic Mathematics** (DVD #2 of 5)	$ 45	100 gm
☐	**Magic Squares** (DVD #3 of 5)	$ 45	100 gm
☐	**The Divine Phi Proportion** (DVD #4 of 5)	$ 45	100 gm
☐	**3-Dimensional Geometry** (DVD #5 0f 5)	$ 45	100 gm

Magic Squares

	Item	Price	Weight
☐	**The Book of Magic Squares, Volume 1** (M. Sqs. of 3, 4, 5, 6, 7)	$ 45	520 gm
☐	**The Book of Magic Squares, Volume 2** (M. Sqs. of 8, 9, 10)	$ 45	520 gm
☐	**The Book of Magic Squares, Volume 3** (M. Sqs. of 11, 12, 16)	$ 45	560 gm
☐	**Join-The-Dots** (adults and children coloring book. Pub:1999, 56 pp)	$ 30	160 gm
☐	**Decal: The Magic Square of 4 X 4 (Jupiter)** (Small, Magenta)	$ 10	20 gm
☐	**Decal: Magic Square 8 X 8 (Mercury)** (Small, Blue Background)	$ 10	20 gm
☐	**Decal: Magic Square 4 X 4 (Jupiter)** (Large, Yellow Background)	$ 20	100 gm
☐	**The Gospel of the Holy Twelve** (Essene Bible w. Magic Sqs. 210pp)	$ 45	410 gm

Vedic Mathematics

	Item	Price	Weight
☐	**The Magic of Nine in Vedic Mathematics.** (Pub:1999, 124pp)	$ 45	390 gm
☐	**The Vedic Mathematics Curriculum for the Global School. Part 1, Digital Sums.** (Pub:2001, 142 pp)	$ 45	460 gm
☐	**Jain Mathemagics Curriculum for the Global School. Part 2. Multiplication.** (pub: 2005, 152pp)	$ 45	480 gm
☐	**Vedic Mathematics for the New Millennium: DVD** (Pub:2002)	$ 45	100 gm
☐	**Dolphin Human Connections:** Jain shows Vedic Maths (2006). 40 mins	$ 50	100 gm
☐	**Vedic Mathematics** by: Bharati Krsna Tirthaji **Soft Cov.** (355pp)	$ 45	460 gm
☐	**Vedic Mathematics** by: ibid. **Hard Cover.** (Pub:1965, 355pp)	$ 55	570 gm
☐	**Calculations @ The Speed Of Light** by Kranti Kiran (pub:2002, 104pp)	$ 45	230 gm

Sacred Geometry

	Item	Price	Weight
☐	**The Book of Phi, Vol 1** (The Living Maths of Nature. Pub:2002, 172pp)	$ 50	550 gm
☐	**In The Next Dimension** (The Book of Phi, Vol 2. Pub:2003, 180pp)	$ 50	580 gm
☐	**Propositions on the Exact value of Pi.** Reddivari Sarva J. Reddy	$ 60	620 gm
☐	**Peter's Projection Map.** (Earth's true cartographic proportions)	$ 25	50 gm
☐	**Dvd: Jonathan Quintin's "Visual Symphonies"** Toroidal Animation	$ 45	100 gm

EarthHeart Jewellery

	Item	Price	Weight
☐	**Platonic Solids:** Crystal Gift Set in wooden camphor boxes	$100	50 gm
☐	**Heart Star:** 24 carat gold plated stellated cuboctahedron	$150	50 gm
☐	**Tantric Star:** 24 carat gold plated Star tetrahedron	$130	50 gm
☐	**Solar Star:** 24 carat gold plated stellated dodecahedron	$150	50 gm

SUB-TOTAL: $ ....

Postage & Handling: $ ....

TOTAL: $ ....

Tax Invoice

Name: ....

Mailing address: ....

City/Town: .... **State**: .... **Zip Code**: ....

E-mail Address: ....

Home phone: ....**Mobile**: ....

Please choose your preferred method of payment:

☐ **Cash, Cheque or Money Order.** Amount: ....
Payable: JAIN F.R.E.E.D.O.M.S.

☐ **PayPal now Available!** Amount: ....
Email: jain@jainmathemagics.com

☐ **Direct Deposit** Amount: ....
Acc Name: JAIN F.R.E.E.D.O.M.S. BSB: 032573 Acc Number: 171268

☐ **(circle) Visa / Mastercard** Amount: ....

Card number..

Name on Card: .... **Exp. date**: ....

AUSTRALIAN POSTAGE: Orders up to $50, add $8. Orders $50 to $120, add $12.

INTERNATIONAL ORDERS: $10 set up fee and add $10 per item. (Approx.)
International Bank Draft in Australian Dollars. PayPal also available + Western Union.
Cheques made out to JAIN F.R.E.E.D.O.M.S.

All Prices are listed in Australian Dollars. **All products available online!**

Date Of Order .... **Dispatched** ....

6	1	8
7	5	3
2	9	4

JAIN F.R.E.E.D.O.M.S. inc 9882763 trading as JAIN MATHEMAGICS
For Research Expressing Essential Data Of Magic Squares
Phone: +61 02 6684 4409 Mobile: +61 0423 583 886 ABN 47 244 872 696
Email: jain@jainmathemagics.com Web: www.jainmathemagics.com
Address: /// Left Bank Rd, Mullumbimby Creek, NSW, 2482, Australia

RAPID MENTAL CALCULATION

VEDIC MATHS for TEENS and ADULTS

MULTIPLICATION

BOOK 2

JAIN 108

from the series:

"JAIN MATHEMAGICS CURRICULUM FOR THE GLOBAL SCHOOL"

JAIN MATHEMAGICS CURRICULUM FOR THE GLOBAL SCHOOL part 2 MULTIPLICATION 2005

"THE FUTURE IN MATHEMATICS IS DOING IT IN YOUR HEAD" (by Jain: *Mathematical* Futurist).

- This is an Educational Workbook for Children and a Teacher's Resource Material for those interested in learning and teaching Rapid Mental Calculation. It increases the child's Memory Power and Confidence.
- This long-waited for book of Ancient Mathematical Short-Cuts excavates many hidden truths and vital properties in the playing field of Numbers.
- There is a rare and unique chapter called "HARMONIC STAIRWAY" from Jain's Dictionary of Numbers, for the first time in print. These choice Mathematical Plums are something that the student stumbles upon. They are working away solving calculations mentally, and when they check their answers in Chapter 2, they are instructed, in the section: DID YOU KNOW, to learn more about that particular number. But these are not any particular numbers, they are Anointed Numbers that our forebears held in high esteem.
- Beautifully and richly illustrated that the graphics alone educate. The student learns about other Families of Numbers, other numerical relatives and mathematical cousins that enrich their understanding of what mathematics is really about. It instills that sense of joy and wonderment that Pythagoras and Baudahayana knew.

- There is no error; it is an infallible bulletproof system, based on Unity Consciousness. The 16 "Threads" or Sutras that solve all known mathematical problems express The Law of Economy and The Path Of Least Resistance.
- As Jain Mathemagics becomes more globally acknowledged, it will help end the generational tyranny that has kept such knowledge in the dusty cupboard. This book is an organic pill that will prevent the slowly encroaching borgificiation of Mathematics.
- Since the turn of the bi-Millennium, there has been a global renaissance in the subject of "Sacred Geometry". You could summarize it as a fascination for the Language of Shape. This book invites seekers to truly incorporate SHAPE into their ability to perform mathematics, to apply this Language of Shape to the next octave of learning, not theoretically, but by practical use of using SHAPE to literally perform mental calculations in seconds.
- Often students come out of school unable to recover from deep mathematical wounds. They are mis-diagnosed as "un-intelligent" or "dumb", given drugs like Ritalin, told they are ADHD (which really means: Attuned Directly to Higher Dimensions). In fact, Dyslexic children are geniuses.
- **UNCOVER, RECOVER, DISCOVER**

LEARN THE
SECRETS OF NATURE
JainMathemagics.com
Jain's 5 Interactive Seminars will help you discover the beauty of numbers in Art, Science, Music and History.
THE ART OF NUMBER
Translating Numbers Into Art and Sacred Symbols
RAPID MENTAL CALCULATION
Vedic Mathematics
Learn how to be a Human Bio-Calculator
MAGIC SQUARES
Translating Numbers Into Exquisite Atomic & Mandala Art
THREE-DIMENSIONAL GEOMETRY
of the 5 Platonic Solids Reveal Secrets of Atomic Structure
DIVINE PROPORTION
Living Mathematics of Nature and the Lost Secrets of the 108 Phi Code
Three Age Groups:
Juniors (9-12)
Teens
Adults
The 5 Day Seminars progress from:
~ Level 1 Introductory
~ Level 2 Advanced
~ Level 3 Teacher Trainer Courses
The Book Of Magic Squares, vol 2, Jain 108
168

The Book Of Magic Squares, vol 2, Jain 108

GOD DESS ENCE
I AM
To Live Life
To Love Life
TO LIVE LOVE

GOD-DESS-ENCE I AM
To Build a Spiritual Platform For Twin Flames To Meet
To Live Life
To Love Life
To Live Love